Studien zur Hochschuldidaktik und zum Lehren und Lernen mit digitalen Medien in der Mathematik und in der Statistik

Herausgegeben von
R. Biehler, Paderborn, Deutschland

Fachbezogene Hochschuldidaktik und das Lehren und Lernen mit digitalen Medien in der Schule, Hochschule und in der Mathematiklehrerbildung sind in ihrer Bedeutung wachsende Felder mathematikdidaktischer Forschung.

Mathematik und Statistik spielt in zahlreichen Studienfächern eine wesentliche Rolle. Hier stellen sich zahlreiche didaktische Herausforderungen und Forschungsfragen, ebenso wie im Mathematikstudium im engeren Sinne und im Mathematikstudium aller Lehrämter. Digitale Medien wie Lern- und Kommunikationsplattformen, multimediale Lehrmaterialien und Werkzeugsoftware (Computeralgebrasysteme, Tabellenkalkulation, dynamische Geometriesoftware, Statistikprogramme) ermöglichen neue Lehr- und Lernformen in der Schule und in der Hochschule.

Die Reihe ist offen für Forschungsarbeiten, insbesondere Dissertationen und Habilitationen, aus diesen Gebieten.

Herausgegeben von
Prof. Dr. Rolf Biehler
Institut für Mathematik, Universität Paderborn, Deutschland

Birgit Griese

Learning Strategies in Engineering Mathematics

Conceptualisation, Development, and Evaluation of MP²-MathePlus

With a foreword by Prof. Dr. Bettina Rösken-Winter

 Springer Spektrum

Birgit Griese
Paderborn University, Germany

Dissertation Ruhr-Universität Bochum, 2016

ISSN 2194-3974   ISSN 2194-3982   (electronic)
Studien zur Hochschuldidaktik und zum Lehren und Lernen mit digitalen Medien in
der Mathematik und in der Statistik
ISBN 978-3-658-17618-1   ISBN 978-3-658-17619-8   (eBook)
DOI 10.1007/978-3-658-17619-8

Library of Congress Control Number: 2017934643

Springer Spektrum
© Springer Fachmedien Wiesbaden GmbH 2017

Printed on acid-free paper

This Springer Spektrum imprint is published by Springer Nature
The registered company is Springer Fachmedien Wiesbaden GmbH
The registered company address is: Abraham-Lincoln-Str. 46, 65189 Wiesbaden, Germany

Foreword

Exploring learning strategies in engineering mathematics has been the essential research focus of the project MP^2-Mathe/Plus, located at Ruhr-Universität Bochum. The project was implemented to help engineering students to succeed in mathematics during their first year of university studies. Connected to the *Servicezentrum Mathematik und Anwendungen* of the faculty of mathematics, the project involved several mathematicians and myself, in my role as mathematics educator.

The PhD work presented in this book is distinguished for several reasons. First, the conceptualization of the project follows a design research approach, bringing together the development of interventions and research on the corresponding effects. Second, all project interventions are documented in detail, informing other faculties about how to compose a successful project for enhancing student learning. Third, the research field of learning strategies has been pushed forward as the construct is researched in depth and from different perspectives. Finally, although the topic is complex as is the research approach, Birgit Griese has managed to unfold the dissertation project in such a way that the reader is well guided throughout his or her journey through the problem identification, the research approach, the design development, the evidence level, and the project evaluation of MP^2-Mathe/Plus.

Supervising the dissertation was distinguished for one reason: The candidate put so much enthusiasm and immense commitment into the project that supporting the PhD thesis was just straightforward.

All that remains now for me is to wish the reader the same delight that I felt during my involvement in this PhD journey.

Bettina Rösken-Winter

Acknowledgement

I am deeply thankful for many things: the support of my parents, my wonderful husband Hannes (T$_{\!E}$X god) who is always there for me, my colleague Michael who never tires of answering my questions, my friends who have endured my frequent absence, and of course my supervisor Bettina who has encouraged me in many ways.

Without you I would not be who I am.

Abstract

MP^2-Math/Plus is a support project for engineering first-years at Ruhr-Universität Bochum. Its aim is to prevent unnecessary drop-out. Conceptualisation and development of the project follow a design research approach (van den Akker, 2013; Gravemeijer & Cobb, 2013). The interventions focus on learning strategies which are collected in a pre-post design with the aid of the LIST questionnaire (Wild & Schiefele, 1994). These and other data are utilised for the evaluation of MP^2-Math/Plus. The results confirm the adaptations of the project procedures in successive cycles, stress the importance of effort and motivation, and assess the success of the project.

Keywords: *secondary-tertiary transition, LIST questionnaire, design research, learning strategies, engineering, mathematics*

MP^2-Mathe/Plus ist ein Unterstützungsprojekt für Studienanfänger in den Ingenieurwissenschaften an der Ruhr-Universität Bochum mit dem Ziel der Verhinderung unnötiger Studienabbrüche. Konzeptualisierung und Entwicklung des Projektes folgen dem Ansatz des Design Research (van den Akker, 2013; Gravemeijer & Cobb, 2013). Die Projektinterventionen fokussieren auf Lernstrategien, die mit Hilfe des LIST-Fragebogens (Wild & Schiefele, 1994) in prä-post Umfragen erfasst wurden. Diese und andere Daten fließen in die Evaluierung von MP^2-Mathe/Plus ein. Die Ergebnisse bestätigen die Anpassungen der Projektmaßnahmen in den aufeinanderfolgenden Zyklen, betonen die Relevanz von Anstrengung und Motivation, und bewerten den Projekterfolg.

Schlagworte: *Übergang Schule - Hochschule, LIST Fragebogen, Design Research, Lernstrategien, Ingenieurwissenschaften, Mathematik*

Contents

1 Introduction: Encountering the Problem

University courses that are (albeit remotely) related to a technical subject such as science or engineering involve a basic knowledge of mathematics. Although this is a subject all students come into close and continual contact with during their schooldays, the passage from secondary to tertiary education is considered particularly problematic regarding the challenges students encounter when being confronted with university mathematics (cf. Gueudet, 2008; Bruder et al., 2010). Students' failure rates in STEM (science, technology, engineering, and mathematics) subjects are alarmingly high in many countries. In Germany, for instance, almost 48% of engineering students fail in their first year of university studies (Heublein, Richter, Schmelzer, & Sommer, 2012). For the USA, Knight, Carlson, and Sullivan (2007) cite retention rates of 47% for the 1990s and an average of 56% for later studies, which is comparable to the German dropout rate. The United Kingdom has much lower dropout rates on the whole (usually single figures), but they are highest for engineering, technology, and computer science[1]. The reasons given are the same across countries, many relate to the difficulty to meet the demands, the extensive preparations and follow-up work needed, and many specifically mention mathematics as the subject that constitutes the main obstacle.

As the challenges are similar everywhere, many universities attempt to support their first-years in various ways. This was also the intention at Ruhr-Universität Bochum when MP²-Math/Plus was introduced[2], the project that lies at the heart of this study. Its objective is to remedy the problem described above, to prevent engineering students from dropping out of their course. The concept idea of MP²-Math/Plus is to do so by advancing students' learning strategies with respect to mathematics. The planning, re-designing and assessing of

---

[1] https://www.timeshighereducation.com/news/uk-has-lowest-drop-out-rate-in-europe/2012400. article, accessed 10/06/2016.

[2] Inside this work, the project name is given in its Americanised form, Math/Plus/Practice, used in previous publications. The German spelling Mathe/Plus or MathePlus only appears in the title of this thesis and on the project website, www.rub.de/matheplus.

MP^2-Math/Plus necessitates a detailed description of the design and re-design of the interventions in the successive project cycles, so the composition of this thesis does not strictly follow the traditional structure of presenting empirical studies. Figure 1.1 relates the respective standard chapters (in the right column) to the titles of the main chapters (in the left column). The shapes in the centre illustrate their interconnections. An orientation figure referring to Figure 1.1 will appear at the beginning of the main chapters in the course of this thesis, and offer additional guidance through its structure.

This short introduction (Chapter 1) is followed by a broad description of the theoretical concepts (Chapter 2) relevant for the research purpose. Here, the theoretical background in the form of models for the transition from secondary to tertiary education in mathematics is inspected (section 2.1.1), the learning of advanced mathematics is conceptualised in various ways (section 2.1.2), together with its obstacles and approaches to overcome them (section 2.1.3). Apart from cognitive aspects of transition, affective and social factors are likewise introduced. This also involves contemplating and structuring the specifities of engineering mathematics (section 2.1.4), other universities' initiatives to support their students (section 2.1.5), and the concrete conditions at Ruhr-Universität Bochum (section 2.1.6). The project focus, learning strategies (section 2.2), is analysed next, and their assessment with the help of the LIST questionnaire is explored. Reviews of the relevant literature are incorporated in the respective sections.

Then, the *Design Research* approach and the research goals are elaborated upon (Chapter 3, which also contains the research questions). This chapter interacts with the two following ones that comprise the project conception and (re-)design (Chapter 4) and the empirical investigations (Chapter 5); the three chapters refer to and induce each other. Chapter 4 elaborates on the project conception (section 4.1) and preconditions for the first project cycle (section 4.2), and it comprises a detailed description of the project procedures (section 4.3). The adaptations decided upon for the ensuing project cycles are presented (section 4.4) and described in detail (section 4.5). Additional modifications for subsequent project cycles are also depicted (section 4.6).

The basis for the changes and the later overall evaluation of MP^2-Math/ Plus is established by diverse empirical analyses (Chapter 5, which covers both methodologies and results). This involves a description of the project applicants (section 5.1), the sample as a whole (section 5.2), students' feedback

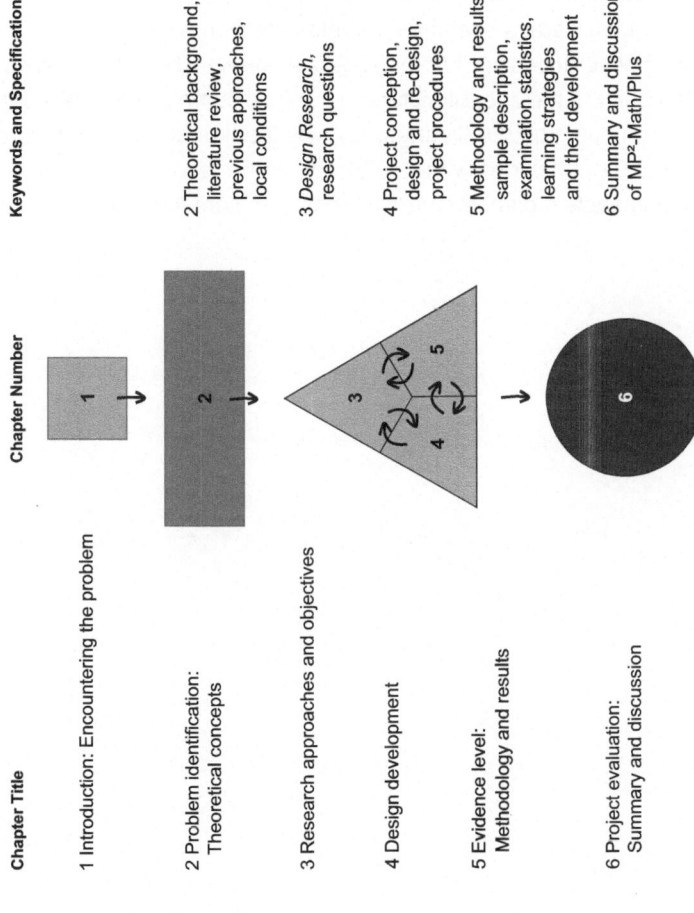

Figure 1.1: Structure of this Thesis with Keywords and Specifications

to the project interventions (section 5.3), and their progress through the most difficult modules of their engineering course (section 5.4). Furthermore, the examination statistics of MP^2-Math/Plus participants and non-participants are explored in detail, including gender aspects (section 5.5). Before the impact of learning strategies on academic success (section 5.7) and their development in the course of the first months at university (section 5.8) are investigated, the factor structure of the data is examined in detail (section 5.6).

Some of the deliberations in Chapter 5 are of a rather technical nature, so the focus is shifted back to the overall picture in the final summary and discussion (Chapter 6), which sums up the findings, reviews them under different aspects (sections 6.1 to 6.3), concludes with a reflection of the research approach and methodology, and casts a glance ahead at possible future implications (section 6.4).

2 Problem Identification: Theoretical Concepts

This chapter contains a literature review and the theoretical background concepts relevant for a support project for engineering first-years in mathematics. This forms the basis for the following chapters covering the research approach (Chapter 3), the design and development of the project (Chapter 4), and the empirical analyses (Chapter 5), see orientation figure below.

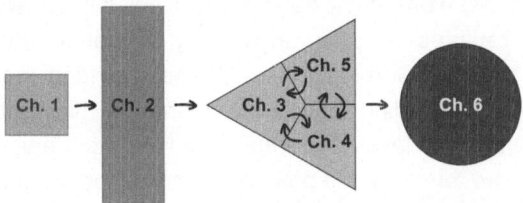

Concerning transition to tertiary mathematics (section 2.1), the material presented here addresses transition processes in general (section 2.1.1), the learning of mathematics at tertiary level (section 2.1.2), and approaches to overcome the problems connected with this (section 2.1.3). The specifities of engineering mathematics (section 2.1.4) are explored as well as concrete projects at other universities (section 2.1.5) and the tangible conditions at RUB (section 2.1.6).

In reference to learning strategies (section 2.2), the assessment of learning behaviour with the help of questionnaires is considered (section 2.2.1), followed by a detailed account of the LIST questionnaire (section 2.2.2).

2.1 Transition to Tertiary Education in Mathematics

A human passing from kindergarten to primary and later to secondary school, and finally to some kind of education at the tertiary level, experiences transition more than once. From a general point of view, all these processes have three stages in common: letting go of what you have known and become accustomed to; getting to know new surroundings, new expectations, and new rules; and finally, getting used to these new concepts. How individuals experience their first transitions in life can influence the way they cope with the next one. This is reason to explore the characteristics of transition processes, in reference to mathematics education.

2.1.1 Characterising transition processes

This section describes the characteristics of transition processes in reference to to mathematics and summarises the specifities for Germany.

Early transition processes

The change from the elementary educational level, kindergarten, with its emphasis on daily routines and play, to primary school is characterised by a formalisation of learning (Grüßing, 2009). Whereas children learn basic mathematical concepts (like adding, taking away, sharing, distributing) in real-life situations at elementary level (in compliance with the ideas of Aebli, 1976), these processes are formalised in primary school, where arithmetic operations are abstracted from situational events (Clements & Sarama, 2007). The passage tends to be smooth, with primary teachers usually having a good idea of what children already know, and with constant reference of abstract operations to the real-life situation behind (Stern, 1997). This transition experience should ideally provide learners with positive feelings of success and independence.

The transition from primary to secondary school can contain obstacles (McGee, Ward, Gibbons, & Harlow, 2003), at least in Germany, where most of the federal states split children into two or three ability groups at the age of 10 or 12. Primary-secondary transition in mathematics is otherwise associated with assimilating to a new (presumably more homogeneous) group of peers, with more teachers for the different subjects, and with potentially higher demands for the achievement of cognitive goals (Reiss, 2009b; Maaz, Baumert, Gresch, & McElvany, 2010; Koch, 2006). Thus the start at a secondary school can be connected with fear of failure (Maaz, Gresch, McElvany, Jonkmann, & Baumert, 2010; Breen & Goldthorpe, 1997). With respect to mathematics, there is a consensus that mathematics at secondary level is more abstract, as it contains a growing number of abstract concepts, such as variables, functions, or infinity (Reiss, 2009a).

Specifities for Germany

The re-orientation that the teaching of mathematics (among other subjects) has undergone in Germany after PISA (*Programme for International Student Assessment*) and TIMSS (formerly *Third International Mathematics and Science Study*, now *Trends in International Mathematics and Science Study*) in the year 2000 (Baumert, 2001; Köller, Baumert, & Bos, 2001) aimed at improving lessons in general. The shock that German students did not rank on top in mathematics led to intensified efforts to help students to understand what mathematics is about (in contrast to just teaching them calculation routines).

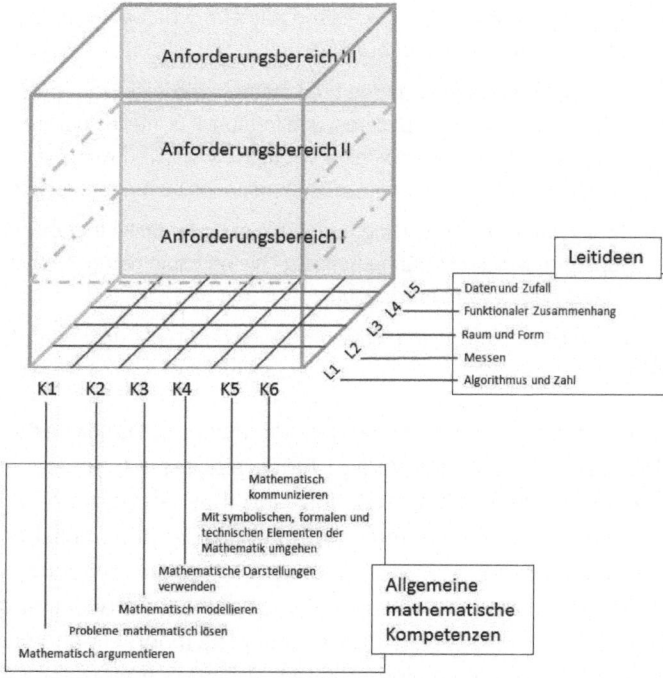

Figure 2.1: Competence Model of Education Standards for Mathematics at Higher Secondary Level, (KMK, 2012, p. 12)

This re-conceptualisation is based on a three-dimensional competence model comprising general mathematical competences (*allgemeine mathematische Kompetenzen*, content dimension), central ideas (*Leitideen*, action dimension), and challenge levels (*Anforderungsbereiche*, challenge dimension), see Figure 2.1.

This rethinking brought the change from stressing what was covered in class, input-orientation, to what students had truly mastered, output-orientation (Bildungsministerium für Bildung und Forschung, 2007), thus putting an emphasis on understanding, visualising, and connecting (Blum, Drüke-Noe, Hartung, & Köller, 2006). Ideally, students should experience mathematics at secondary level according to the three fundamental principles described by Winter (1996):

(G1) To recognize and to understand phenomena in the world around us. (This principle recognizes the role of mathematics in acquiring important knowledge of our world.)

(G2) To learn about and to understand mathematical issues represented in language, symbols, pictures, and formulas as intellectual creations, to recognize mathematics as a deductively ordered world of its own kind. (This principle recognizes mathematics as a rigorous science.)

(G3) To acquire problem-solving (heuristic) skills for tasks that extend beyond the domain of mathematics. (This principle recognizes mathematics as a school of thought.)

Winter (1996) – English translation and explanations according to H.-W. Henn (2003, p. 72).

In relation to the teaching of calculus, for instance, Danckwerts and Vogel (2006) discuss the tensions between those placing G1, real-world applications, in the centre of teaching calculus, and those preferring G2, concepts and interconnections. They summarise the criticism of established teaching approaches (Danckwerts & Vogel, 2006, p. 11) as the "tendency to neglect the fundamental principles G1 (applications) and G3 (heuristic strategies) in favour of the fundamental principle G2 (calculus as theory)"[1], and commiserate the reduction of G2 to calculation routines. Although their reflections refer to

---

[1] „Tendenz, die Grunderfahrung G1 (Anwendungen) und G3 (heuristisches Arbeiten) zugunsten der Grunderfahrung G2 (Analysis als Theorie) zu vernachlässigen", translation by author.

the teaching of calculus (which is the central issue in the advanced mathematics classroom), their descriptions fit other areas of teaching mathematics as well. In theory, Winter's triad of fundamental principles forms a perfect balance, as G3 (heuristic strategies) emphasises systematic testing, relating features to each other, and reasoning, and thus complements and deepens G1 and G2, which are regarded (respectively) as thought-out solutions for meaningful real-world problems and a developing network of interconnected concepts. In practise, however, there is severe doubt if these aspirations are being met for the majority of German students (Frey, Asseburg, Carstensen, Ehmke, & Blum, 2007; Frey, Asseburg, Ehmke, & Blum, 2008). Frey and colleagues present their findings of 15-year-olds' competencies (in the sense of a flexible ability to apply knowledge and skills to problems in a meaningful context). This (PISA) approach allows comparisons between countries and is detached from short-term rote learning, thus perfectly adequate in connection with the fundamental principles described above. The results place Germany in the middle range of OECD (Organisation for Economic Co-operation and Development, currently 34 member states, 21 from Europe) states. The findings show, among other results, that girls are significantly less competent in mathematics than boys. This point particularly implies need for further development in terms of teaching and instruction concepts, as does the fact that the dependency of educational success on social background is still remarkable, although it has slightly weakened in the three years since the previous study. Another important result includes that almost 20% of the teenagers have reached no more than the lowest competence level in mathematics, and that the variance of competence levels is comparably big. This last fact, in combination with the circumstance that the percentage of a generation that enrols for university courses has increased considerably in Germany in the last years from under 20% in the 1980s, to 34.4% in 2007, to 51.3% in 2013[2] supports the assumption that a notable percentage of students aspiring to university have not experienced mathematics as a field for intellectual involvement that requires active pursuit of understanding and cognitive processing (as Fischer, Heinze, and Wagner (2009) summarise in their paper), they rather see mathematics as an unconnected collection of cryptic formulas, unintelligible rules, presented via abstract formalism.

---

[2] See Bundesministerium für Bildung und Forschung (n.d., Tabelle 1.9.3).

Secondary-tertiary transition

In relation to other subjects, the gap between school and university mathematics seems extremely high and causes difficulties for students taking mathematics courses; Engelbrecht (2010) even describes first-year experiences at university as "traumatic" (p. 143). Although school mathematics itself is regarded as a tough, even polarising subject (H.-W. Henn & Kaiser, 2001), it is generally accepted that university mathematics is even tougher (Dreyfus, 1995; Zucker, 1996). The dramatic character is depicted in the word "abstraction shock", used by some authors since there is a notable difference between mathematics taught at school and at university.

Clark and Lovric (2008) understand secondary-tertiary transition in mathematics as a rite of passage (a well-studied anthropological concept), an initiation into a new world. As such, they describe the three stages of transition from high school to university as follows:

- separation (from high school); this stage takes place while students are still in high school, and includes anticipation of forthcoming university life;
- liminal phase (from high school to university) includes the end of high school, the time between high school and university, and the start of the first year at university;
- incorporation (into university) includes, roughly, first year at university (Clark & Lovric, 2008, p. 35)

In accordance with theories for rites of passage, Clark and Lovric realise the fact that secondary-tertiary transition is a "stressful, demanding, life-changing experience" (Clark & Lovric, 2008, p. 29), that it "involves both body and mind" (Clark & Lovric, 2009, p. 759), and that it needs a supportive environment, meaning "the totality of the contexts involved (social, psychological, cognitive etc.)" (Clark & Lovric, 2009, p. 759). This stresses the complexity of the transition process, particularly the notion that it is not only cognitive aspects that pose the obstacles – although studies often elaborate on cognitive difficulties and conceptual obstacles experienced by students in reference to specific content, such as aspects of functions (Attorps, Björk, Radic, & Viirmann, 2013; Viirmann, 2013; Winsløw, 2013), of linear algebra (Jaworski, Treffert-Thomas, & Bartsch, 2009; Hausberger, 2013), or proof (Hoffkamp, Schnieder, & Paravicini, 2013;

Selden & Selden, 2005; Moore, 1994). The way mathematics is communicated at university is also an area of intense research, cf. Artigue, Batanero, and Kent (2007).

There is a growing interest in investigating the various barriers that students encounter when starting a university course involving mathematics. The working groups on *University Mathematics Education* at the conferences of ERME (European Society for Research in Mathematics Education) discussing this subject reflect this trend: there were 21 papers in 2011, 23 in 2013, and 35 in 2015 submitted to and accepted by the group dedicated to the specifities of mathematical education at tertiary level, cf. CERME 7 (2011), CERME 8 (2013), and CERME 9 (2015).

The challenge of comprehending (and influencing) how the learning of mathematics at university works is therefore often addressed by the use of cognitive development theories, as we will do in detail in the first part of section 2.1.2. But other aspects play a role, too; that is why we will explore non-cognitive aspects later in section 2.1.2, such as motivation, self-regulation, attribution theories, and particularly learning strategies in section 2.2.

2.1.2 Learning mathematics at university

Learning mathematics at tertiary level is a complex matter. For our project seizing on elements of *Design Research*, we are looking for theories that can be applied to higher mathematics, that describe learning processes adequately in their complexity, and that comprise cognitive as well as affective and motivational aspects. The last point is imperative as, apart from researching the conditions that support or hinder academic success in mathematics for engineering students, the focus lies on the interventions themselves, on approaching preconditions that are changeable.

In this chapter, we will subsequently look at how mathematics is taught and learned at university, and at the differences and commonalities between school mathematics and university mathematics. This leads to the cognitive stages essential for the development of abstract mathematical thought and concepts, to theories on advanced thinking and eventually to the non-cognitive concepts involved. Each section closes with a short retrospect, and finally chapter 2.1.2 ends with a conclusion of which theories are relevant for the research perspective at hand.

Mathematics at secondary and tertiary level

In accordance with the special position mathematics occupies both at school and at university, most researchers introduce a category to describe the difficulties stemming from the high level of abstraction and complexity typical of this subject. Various terms are used: de Guzmán, Hodgson, Robert, and Villani (1998, p. 747)[3] name this category "epistemological and cognitive", Rach and Heinze (2013, p. 123) refer to it as "the aspect of subject content in mathematics and its character"[4], and Gueudet (2008, p. 237) calls it "individual difficulties". The choice of terms here strictly distinguishes between individual lack of effort or perception, and "institutional practices" (Gueudet, 2008, p. 247) that entail "limited (...) and poorly connected" (Gueudet, 2008, p. 246) mathematical organisations unsuitable for the extensions needed at university level. Various work is dedicated to describing different facets of this category, e.g. the formality of mathematics, the stringent way of thinking, and the systematic structure. In this respect, the skills and knowledge shown by school leavers are often found wanting, like in Bruder et al. (2010), Hoyles, Newman, and Noss (2001), or Mündemann, Fröhlich, Ioffe, and Krebs (2016).

Rach (2014) gives a comprehensive overview that contrasts mathematical subject matter at secondary and tertiary level in five categories, see Table 2.1. She proceeds from the understanding that school mathematics and scientific mathematics have different goals: Whereas at school, the aim is to generally educate students and to enable them to solve text problems with the help of mathematics (in a world where concepts possess a concrete, or at minimum a symbolic representation), at university the aim is to introduce mathematical theories in a formal-axiomatic world[5]. Concerning proofs, apart from the fact that they are the central activity in university mathematics, Rach differentiates between dominant features and types on the one hand, and formalisation on the other. In each of these two subcategories, university mathematics is stricter,

---

[3] In this study, engineering students tended to see their passage from school to university as less difficult than students with mathematics major or aspiring teachers, although it was still "24 out of 118 (20%)" (de Guzmán et al., 1998, p. 749) who agreed or totally agreed to finding the transition to university mathematics difficult.

[4] "der Aspekt (1) des Lerninhalts Mathematik und dessen Charakter", Rach and Heinze (2013, p. 123), translation by author.

[5] These terms refer to the theory of the three worlds of mathematics, cf. Gray and Tall (2001) and Tall (2004).

Table 2.1: Comparison of Characteristics of Mathematical Subject-Matter at Secondary and Tertiary Level, Rach (2014, p. 79), shortened and translated by the author

Category	School Mathematics	(Scientific) Mathematics, first-year course at university
Goal	general education, esp. solving text problems with mathematical knowledge	learning about scientific mathematics, esp. building mathematical theories
Mathematical thought processes	conceptual-embodied world / proceptual-symbolic world	formal-axiomatic world
Proofs: activity	one of many mathematical activities	central mathematical activity
Proofs: dominant feature and types	mostly explanation, partly verification; experimental or pre-formal proofs	more verification, additionally communication; formal-deductive proofs
Proofs: formalisation	low level of mathematical notation	high level of mathematical notation
Concept formation: objects and axiomatic system	objects refer to concrete real objects, embedded into contextual axiomatics	objects describe mental constructs, embedded into formal axiomatics
Concept formation: kind of description, level of abstraction	description via specific representatives	abstraction: description via characteristic features
Concept formation: level of formalisation	low level of formalisation	high level of mathematical formalisation

more formal, and more abstract (in the sense of further removed from examples and experiments).

While Rach contraposes school mathematics and university mathematics, Dreyfus (1991) sees the differences as more of a continuum, where generalizing ("To generalize is to derive or induce from particulars, to identify commonalities, to expand domains of validity", p. 35), synthesizing ("To synthesize means to combine or compose parts in such a way that they form a whole, an entity.", p. 35), and abstracting (closely connected to generalizing) are gaining more and more dominance. The notion of the learner experiencing different worlds of mathematics, as described by Gray and Tall (Gray & Tall, 2001; Tall, 2004), is incorporated here. They developed the theory that there are basically three different kinds of mathematical *objects*, each representing a mathematical world that has to explored and experienced in order to reach understanding:

- those that arise through *empirical abstraction* (in the sense of Piaget[6]) by which is meant the study of *objects* to discover their properties
- those that arise from what Piaget termed *pseudo-empirical abstraction* from focusing on *actions* (such as counting) that are symbolised and mentally compressed as *concepts* (such as number)
- those that arise from the study of *properties*, and the logical deductions that follow from these, found in the modern formalist approach to mathematics.

(Tall, 2004, p. 29)

Figure 2.2 illustrates the concept of the three worlds: There is an overlap between the first two worlds, the *embodied* and the *symbolic world*. Both belong to the two lower levels of practical and theoretical mathematics, whereas the second needs the first to build concepts through perception and action. The third *(axiomatic formal)* world is based on the two first and is the only one to be attributed to the highest level of formal mathematics – and thus to university mathematics, see Table 2.1, third row. The step from the embodied world to the symbolic world is regarded as smaller than the step to the axiomatic formal world, as the observation of objects is followed naturally by the handling of objects – but abstractly thinking about objects is further away.

---

[6] For more information on Piaget's theories on the development of cognitive abilities, see next section.

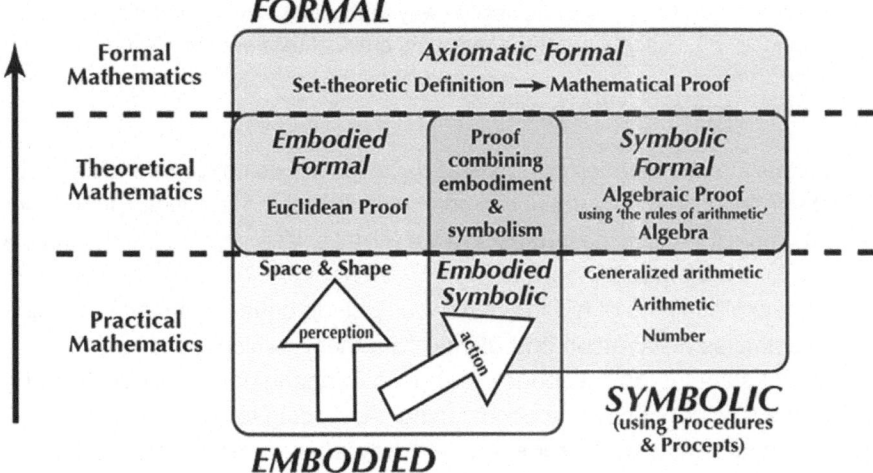

Figure 2.2: The Three Worlds of Mathematics, https://homepages.warwick.ac
.uk/staff/David.Tall/themes/three-worlds.html

The question remains, however, of how the transition from one world to the next is to be accomplished. In university mathematics contexts, it is not unusual to present definitions and theorems in an axiomatic way, disregarding the demand "that students be given a feeling for the long history of magnificent failures" (von Glasersfeld, 1991, p. 179).

In sum, it has become clear that there are notable differences between school and university mathematics, in the areas of abstraction, formality, and concepts. These pose obstacles not easy to overcome.

Cognitive development

The development of understanding and cognitive abilities is of high importance for the learning of higher mathematics. In this section, we will look at the established theories of human behaviour and changes therein, i.e. learning. These theories aim at explaining cognitive processes via development models. When von Glasersfeld states, referring to his lifelong research focus, that

> Constructivism [...] is *one* possible way of thinking. It is a *model* – and models, no matter how useful they might prove, must never be claimed to be 'true'.
>
> (von Glasersfeld, 1991, p. 169)

this means that the theories on cognitive development do not come in categories of truth or deception, but appear in a continuum between viability and inutility for a given context. This is the perspective we will take when sorting the theoretical models for our purposes.

Early explanations of human behaviour include behaviorism (whose main representatives are Watson and Skinner[7]), a theory which focuses on simple models of stimulus and response and aims at changing behaviour patterns by using classical or operant conditioning. According to this way of thinking, an individual reacts in a predetermined way after experiencing a stimulus, which can be anything from a smell to a verbal utterance. The response can be influenced by combining different stimuli (classical conditioning) or by reinforcing wanted behaviour, sometimes in combination with punishing unwanted actions (operant conditioning). When applied to complex learning scenarios, behaviorism shows its limitation, true to Eisenberg's statement that "models of learning are either too specific or too general" (1991, p. 142): The teacher is seen as the centre of the learning situation, as he or she decides which behaviour is wrong and which is right, meaning ignoring or punishing the one and praising the other. This theory does not consider internal mental processes of building understanding, nor does it see the learner as taking an active part in the learning process.

In relation to learning mathematics at tertiary level, this approach is inappropriate in more than one way:

- There is no teacher / lecturer to take complete control of every individual student's learning progress when many hundred of them are affected.

- Mathematical learning at this level implies self-reliant work.

- The nature of mathematics at university (in the sense of the axiomatic formal world) requires understanding, not repeating.

---

[7] Watson (1913), Skinner (2002)

Therefore, behaviorism is inappropriate to describe a modification of human behaviour that stresses the mental processes and incorporates the idea of the learner taking an active part, of learning at a higher level.

Table 2.2: Piaget's Four Phases of Cognitive Development

Development Stage	Age
1 Sensorimotor Stage	0 to 2 years
2 Pre-operational Stage	2 to 7 years
3 Concrete Operational Stage	7 to 11 years
4 Formal Operations Stage	11 to 16 years and onwards

Epistemology, the theory of knowledge, is more suitable, it investigates how human beings develop knowledge and skills, apposite to mathematics at tertiary level. Piaget, one of the earliest representatives of non-behaviorism, saw four phases of cognitive development (Piaget, 1973), as shown in Table 2.2. At the core of Piaget's theory of cognitive development lies the notion that children build an understanding of the world as they experience it through perception and reflection. Whenever a child sees a discrepancy between its reconstruction of the world and a new perception, he or she will adapt their understanding to accommodate it. This has lead to the belief that children need a stimulative and inspiring environment in order to develop cognitively.

In the first stage, young children (before the age of having acquired language) experience the world through sensoric perceptions and discover how to use their motoric abilities. At the age of around two years, children usually discover language and thus reach the pre-operational stage where they play in the sense of pretending things to take a symbolic meaning (e.g. a box becomes a plane). At this stage, children are aware of objects without having them in front of them, but they find it difficult to occupy another's point of view. The later part of the pre-operational phase is characterised by aspiring after knowledge, particularly asking questions as to why things are the way they are. Some of Piaget's famous experiments for determining which stage a child has reached have a mathematical background as they refer to deciding between *more* and *less*, e.g. judging if an amount of liquid stays the same when poured into a narrower but higher container, if a row contains more blocks when they are spread more widely, or deducing the relation between two objects *A* and

C when $A > B$ and $B > C$ is given (transitivity). So, from a mathematical point of view, the pre-operational stage contains the beginning of mathematical reasoning, if only from acquiring misconceptions whose overcoming will later mark the transition to a new phase. The third stage, the concrete operational stage, is even more important in this respect: From the age of about seven years on, children develop inductive logical thinking (although the concept of transitivity may still pose problems), e.g. they understand conservation (that an amount of liquid stays the same no matter the shape of the container). They are more and more able to judge hypothetical situations without being exposed to them, and they start to use trial and error in a systematic way to solve concrete problems. In the formal operations stage, abstract concepts without relation to reality can be grasped, and symbols are used in a logical way. Trial and error is employed more systematically, which marks the beginning of problem-solving in a mathematical sense. Metacognition, the ability to reflect about thinking, is possible.

Although he saw close relations between biological and cognitive deve-lopment, Piaget himself believed his theories could be applied to scientific knowledge as well – which Gray and Tall have done successfully (Gray & Tall, 1994; Tall, 2004) in their theory of the three worlds, see above. For the tertiary level, only the third and fourth stage are relevant, because, in summary, children begin to think logically in the concrete operational stage, but are in need of practical assistance and support. In the formal operational stage, they can think logically without practical help and can develop abstract thought[8]. University students have certainly passed the age of the concrete operational stage, but might profit from practical help when trying to understand complex formal concepts. Piaget stresses that the transition from one stage to the next forms a cycle of repeated actions, observations, and reflections, new actions with slightly different starting positions, new observations and consequently new reflections. He sees the learners as actively influencing the arrangements and observing the outcomes. This idea of learning by acting, observing, and reflecting corresponds with constructivism (see above) - as David Tall summarised as early as 1991: "The active participation in thinking is essential for the personal construction of meaningful concepts. Students

---

[8] It is an interesting issue to investigate if students of mathematics at university have actually reached the *Formal Operations Stage* – Leongson and Limjap (2003) found they had not.

need to be challenged to face the cognitive reconstruction explicitly, through conjecture and debate, through problem-solving [...]" (Tall, 1991b, p. 258f.). The same postulation of actively involving leaners in the learning process is highlighted by von Glasersfeld (1991, p. 175) who states that it "is an illusion that there is knowledge in textbooks or documents. [...] Texts *contain* neither meaning nor knowledge – they are a scaffolding on which readers can build their own interpretation".

Learning in this constructivist sense involves much social interaction, particularly between the teacher and the student. This needs knowledge and understanding for the point of view with which the student regards his or her environment, in summary "an almost infinitely flexible mind" (von Glasersfeld, 1991, p. 178). In detail, the implications of constructivism on a theory of learning are described by von Glasersfeld (1991, p. 177f.) as, among others:

(3) If teachers want to modify a student's concepts and conceptual structures, they have to try and build up a model of the particular student's own thinking. Models of students' thinking can of course be generalised, but before assuming that a student fits the general pattern one should have some solid evidence that this is a viable assumption in the particular case. It should never be assumed that students' ways of thinking are simple or transparent. [...]

(6) Successful thinking is far more important than "correct" answers. Successful thinking should be rewarded even if it was based on unacceptable premises.

(7) To understand and appreciate students' thinking, the teacher must have an almost infinitely flexible mind (because students sometimes start from premises that seem inconceivable to teachers).

All in all, cognitive development, particularly in reference to higher mathematics, is a complex process requiring skilled teachers or lecturers, purposeful interaction, and time.

Advanced Mathematical Thinking

The necessity to inquire specifically into the processes and obstacles of advanced mathematical issues was particularly acknowledged in 1985, when a PME (International Group for the Psychology of Mathematics) Working Group on *Advanced Mathematical Thinking* was established. This resulted, among various other work, in a seminal volume by Tall (1991a) some years later, followed by a special journal issue by Dreyfus (1995) as an update. Much of the research in this area is dedicated to mathematics at tertiary level, but there is general agreement that advanced thinking should and must start at lower levels, as Selden and Selden (2005) assessed appropriately. Thus, in this context, *advanced* is not necessarily applied to *mathematics*, but to *thinking*.

A starting point is the fact that at university, mathematics is presented to the learners via the traditional deductive sequence of *definition – theorem – proof (– application)*, which does not reflect the nature of building mathematical knowledge, which happens "through trial and error, through partially correct (and partially wrong) statements, through intuitive formulations in which loose terms and imprecisions have intentionally been introduced, through drawings that try to visually present parts of the mathematical structures being thought about, through dynamic changes being made to these drawings, etc." (Dreyfus, 1991, p. 27). But as "the constructivist teacher should never present a solution as the only one" (von Glasersfeld, 1991, p. 179), other courses of action are imperative. To re-enact the process of finding connections and describing characteristics, Tall lists

- the participation of the student in the process of mathematical thinking through an active process of "scientific debate", rather than passive receipt of preorganized theory,

- the direct confrontation of the student with conflict which occurs in developing new theoretical constructs, to help them reflect on the problem and build a new, more coherent, cognitive structure.

- the building up of appropriate intuitive foundations for the advanced mathematical concepts, through an approach which balances cognitive growth and an appreciation of logical development.

- the use of visualization, particularly utilizing a computer, to give the student an overall view of concepts and enabling more versatile methods of handling the information,

- the use of programming to cause the student to think through mathematical processes in a way which can be encapsulated by reflective abstraction.

Tall (1991a, p. xiv-xv)

Much of this list is still up to date today, 25 years later, and can well suit as a guideline on which to model teaching concepts in mathematics, both at school and at university level. The suggested techniques are focused on activation of the learner, but the overall implication remains that the teaching concepts must originate from the teacher or lecturer, e.g. in not (or at least not exclusively) presenting content in a preorganised way that expects the learner to listen passively, or by setting exercises to force the learner to structure a mathematical process. This is in keeping with Piaget's theories on how knowledge is built in general. A selection of how mathematical knowledge is built specifically is presented in the following sections.

Procept Theory

The term *procept* was introduced by Gray and Tall (1991, 1994) in order to describe the twofold characteristic of mathematical symbols like $3 + 4$, which at the same time denotes the process of adding two numbers and the concept of the sum it results in. The authors define:

> An *elementary procept* is the amalgam of three components: a *process* which produces a mathematical *object*, and a *symbol* which is used to represent either process or object. [...]
> A *procept* consists of a collection of *elementary procepts* which have the same object.
> (Gray & Tall, 1994, p. 120)

Although many examples of procepts are taken from the area of elementary mathematics, the theory can be applied to more advanced concepts. Flexibility in the use of processes and representations for a given concept is the key to true understanding of mathematics at all levels. For example, the symbol $f(x) = x^2 + 4$ stands for the process of mapping x to $x^2 + 4$ and for the concept of the resulting quadratic function, which in turn has specific features and visual representations. A learner shows his or her expertise in flexibly switching between the process, the concept, its visualisations and the connected features of an object. Gray and Tall have incorporated their thoughts on this matter into the wider theory of the three worlds of mathematics, see above and Tall (2004). And the classification fits indeed: Studying an object in order to discover its properties (the first of the three worlds) corresponds to the object side of a procept, whereas handling and manipulating an object in order to further unveil its properties (the second world) focuses on the process aspect. When these explorations are conducted mentally and formally (i.e. a certain flexibility is reached in the use of the different aspects of a procept), the level of the third world is reached.

Procept theory allows the description and classification of mathematical comprehension problems, as, depending on the misconception shown, the lack of flexibility or variety can be diagnosed – and subsequently ameliorated. Like all theories on mathematical learning, it draws on stimulating exercises and the discussion of their solutions, thus stressing the importance of the teaching concept behind a university course and the design of its central tasks.

APOS Theory

The inventors of APOS theory, Dubinsky and McDonald (2001), view theory in general from a practical perspective, with regard to the help it can provide in mathematics education, substantiated in the six features as shown in Table 2.3. All these features refer to putting a theory in use in order to understand and improve teaching and learning processes. It has to be admitted, though, that this point of view is no exception: Most theorists develop an theory with the aim of predicting future outcomes, of explaining phenomena, or of helping to organise thinking and communication. But in the development of APOS, these guidelines were expressedly the starting point.

Table 2.3: Six Features for Models and Theories in Mathematics Education, according to Dubinsky and McDonald (2001, p. 275)

1	Support prediction
2	Be applicable to a broad range of phenomena
3	Serve as a tool for analyzing data
4	Have explanatory power
5	Help organize one's thinking about complex, interrelated phenomena
6	Provide a language for communication of ideas about learning that go beyond superficial descriptions

APOS theory was introduced by Dubinsky and McDonald (2001) in order to bring together the understanding of learning processes and the observations

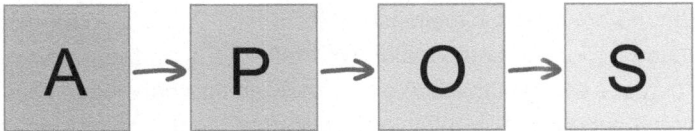

Figure 2.3: Stages of APOS Theory in Hierarchical Order

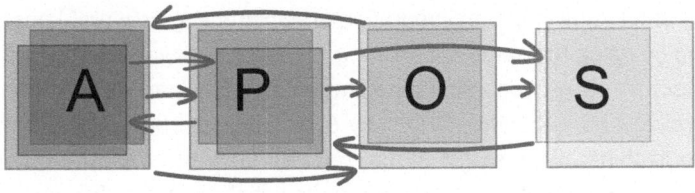

Figure 2.4: APOS Theory Accommodating More Sophisticated Sequences

made when teaching undergraduate students in mathematics. The concept of learning behind APOS theory is again constructivism (see above), meaning that individual learners have to (re-)build the constructs in order to comprehend them. This idea of learning comprises challenging problems, time to think and communications with others (an instructor or fellow learners). APOS has gained many supporters (Arnon et al., 2014), one reason for this is the stringent adherence to the six features from Table 2.3.

APOS theory focuses on the mental processes of an individual when learning mathematics. The hypothesis behind APOS theory is that individuals construct "mental *actions*, *processes*, and *objects*" and organize them in "*schemas* to make sense of the situations and solve the problem" (Dubinsky & McDonald, 2001, p. 276), implying the acronym APOS. Ideally, these four stages of building understanding are passed in linear order by the learners (Figure 2.3), but in reality, the learning process is often more sophisticated (Figure 2.4).

The elements of APOS theory are defined as follows: When an individual is in the act of comprehending a mathematical concept, first he or she starts with performing *actions* on mathematical objects, e.g. the algebraic transformations on a differential equation. At this first stage, these *actions* are perceived as "essentially external and as requiring [...] step-by-step instructions on how to perform the operation" (Dubinsky & McDonald, 2001, p. 276). With these accumulated *actions* in combination with reflections on them, the individual forms a *process* which in this context means "an internal mental construction [...] which the individual can think of as performing the same kind of action, but no longer with the need of external stimuli" (Dubinsky & McDonald, 2001, p. 276). In the example of differential equations, this would mean that the learner now combines and reverses *actions*, in the example above this might involve including other knowledge on integration or constructing new problems that can be solved by using the same *actions*, e.g. separation of variables or variation of parameters. An *object* describes the stage when the learner "becomes aware of the process as a totality" (Dubinsky & McDonald, 2001, p. 277) and does not have to think about single steps, but can view the transformations as a whole and conjecture of how variations in the initial equation affect the outcome. The *schema* finally denominates the individual learner's compilation of *actions*, *processes* and *objects* as well as the network that connects them, in the form of general principles and in the form of concrete relations. In the example above this would comprise a universal ability to solve (certain kinds of) differential equations. APOS' origin in Piaget's phases of cognitive development is obvious; the advantage of APOS is the specifity for mathematics, which Piaget's theory does not feature. As such, APOS is a generally accepted theory to classify cognitive mathematical processes, although there is criticism (Tall, 1999) that it may not apply very well to very advanced or very formal concepts.

So, apart from the conceptual details, APOS theory sees learning as an activity of the learner, and the teacher / lecturer / tutor as supporting this activity by establishing a general set-up that facilitates movement from one stage to another.

Concept Image and Concept Definition

Teaching mathematics in the traditional, deductive way involves starting from definitions of the concepts that are to be covered. Most of the time, these definitions are formal, and desirably, according to Vinner (1991, p. 65f.), they should be "minimal" (not containing parts that can be inferred from others), "elegant" (which is a question of taste, undoubtedly, but most mathematicians will consider short, condensed definitions as more elegant than longer, elaborated ones), and (more or less obviously) "arbitrary". Even when a definition does not stand at the beginning of the process of introducing a concept, but at the end of a phase of exploring objects and their characteristics, the formal definition is expected to be the reference to turn to when arguing for or against certain properties. Tall and Vinner (1981) found that this was often not the case, though, because the mental image formed in the mind of the learner takes precedence over the formal definition. Thus, intuitive thinking via the concept image prompts the answers, without reference to the concept definition. Particularly with concepts such as functions, continuity, and limits, Tall and Vinner (1981) as well as Eisenberg (1991) showed that even good students have problems forming concept images that will hold in non-standard situations.

RBC Theory / Abstraction In Context

On the quest of not only a theoretical framework, but of what can be observed to determine the level of understanding a student has reached, and what can be done to foster (deep) understanding, Dreyfus (2012) developed a theory that is based on "three observable epistemic actions: Recognizing, Building-with and Constructing" (p. 1) and describes how abstract mathematical concepts can be learned in classroom situations. It postulates the idea of the learner as actively and responsibly engaging in the learning process in an inquiry-encouraging environment. Based on the works by Freudenthal (1991), Davydov (1990) and thus Vygotsky (1978), the three components are described as follows, complemented by a fourth, consolidation:

> Recognizing refers to the learner realizing that a specific previous knowledge construct is relevant in the situation at hand. Building-with comprises the combination of recognized constructs, in order to achieve a localized goal such as the actualization of a strategy, a justification or the solution of a problem. The model suggests constructing as the central epistemic action of mathematical abstraction. Constructing consists of assembling and integrating previous constructs by vertical mathematization to produce a new construct. It refers to the first time the new construct is expressed or used by the learner. This definition of constructing does not imply that the learner has acquired the new construct once and forever; the learner may not even be fully aware of the new construct, and the learner's construct is often fragile and context dependent. Constructing does not refer to the construct becoming freely and flexibly available to the learner. Becoming freely and flexibly available pertains to consolidation.

Consolidation is a never-ending process through which students become aware of their constructs, the use of the constructs becomes more immediate and self-evident, the students' confidence in using the construct increases, and the students demonstrate more and more flexibility in using the construct.

(Dreyfus, 2012, p. 3f.)

The ensuing theory is consequently abbreviated RBC+C, and often focuses on the highly relevant C-processes, which can be further differentiated into "different modes of thinking: numerical (C1), algebraic (C2), analytic (C3), and visual (C4) (Dreyfus, 2012, p. 9). Together with his co-authors Hershkowitz and Schwarz, Dreyfus described in detail this theoretical-methodological framework whose long name is "dynamically nested epistemic actions model of abstraction in context" (see Hershkowitz, Schwarz, & Dreyfus, 2001; Schwarz, Dreyfus, & Hershkowitz, 2009). Due to the broad observational basis, the framework covers numerous aspects of the process of forming abstract mathematical concepts. First, it is characteristic for abstract mathematical learning processes that the learner feels the need for a new construct, this implies an active learner and a (teacher- or lecturer-created) challenging learning situation. Second, there is often a moment of enlightenment (or aha) in which the learner accepts a new concept into his or her existing network of mathematical concepts. Third, newly-accepted concepts are not necessarily completely correct. RBC+C allows for studying partially correct or wrong concepts and for identifying the nature of the misconception. Fourth, social interaction plays a major role in the forming of new abstract concepts as "Abstraction often takes place in interacting groups of students" (Dreyfus, 2012, p. 12). Fifth, the use of technology to help test new concepts for their validity, is another facet. All these aspects are covered in the RBC+C framework (also called AiC, short for abstraction in context). As AiC was developed observing students interacting, i.e. used as a methodological approach, the researchers were well aware of the dual nature of the framework as a theory and a methodology, and many have found this fact an asset to their own research (see Bikner-Ahsbahs et al., 2010; Wood, Williams, & McNeal, 2006).

The universal applicability as well as the emphasis on learner activities and how they should be guided makes this framework a promising candidate for the purpose we have in mind for the project of supporting engineering students in mathematics. It also suits the overall concept of *Design Research* in its two-footed balance of theory and method, see section 3.1.

In recapitulation, all models presented for explaining progress in mathematical thinking, have the common feature that they require learners and teachers or lecturers to interact, that learners have to adapt new concepts to already acquired ones, and that this process is often not strictly linear, but rather characterised by misconceptions, inconsistencies, and intellectual contention.

Non-cognitive aspects

Apart from the agreement on cognitive difficulties, the research approaches differ slightly when describing obstacles in the transition from secondary to tertiary education in mathematics in regard to aspects not related to cognition: de Guzmán et al. (1998) call the remaining categories "sociological and cultural" and "didactical" (p. 749), Gueudet (2008) chooses the labels "social" and "institutional" (p. 237), and Rach and Heinze (2013) classify them as referring to institutional frame conditions of university education and individual requirements respectively determinants of learning mathematics at university[9]. Inverted, they mean roughly the same, i.e. what Rach and Heinze see as characteristics of the individual are seen as features of society by de Guzmán et al. and Gueudet. All researchers identify the learning conditions at tertiary level as a problem area, although they place themselves at different positions of the continuum between the extremes *researcher* and *university teacher*, between theory and practise.

An interesting perspective is presented by Gueudet (2008), who focuses on the interrelations of difficulties observed, possible views of transition and the resulting didactical actions. Thus, when describing difficulties, she combines the observation of subject-specific difficulties (like abstract thinking when conducting proofs) with social problems (like regarding training for university lecturers as unnecessary) and institutional obstacles (like deficient textbooks). This comprehensive view yields a more precise picture of the mechanisms involved. The cause of the difficulties experienced varies, depending if the focus is on the students' or the university teachers' point of view, as (de Guzmán et

---

[9] „Aspekte [...] der institutionellen Rahmenbedingungen der Hochschulbildung sowie (3) der individuellen Voraussetzungen bzw. Determinanten des Mathematiklernens an der Hochschule", p. 123.

Table 2.4: Causes of Difficulties in the Learning of Mathematics as Seen by Students and University Teachers, cf. de Guzmán et al. (1998)

Students	University Teachers
way teachers present mathematics at the university level	lack of interest in mathematics
changes in the mathematical ways of thinking	lack of prerequisite knowledge
lack of appropriate tools to learn mathematics	deficient learning style, e.g. lack of organisation or autonomy

al., 1998) have found, see Table 2.4. University teachers tend to attribute the difficulties exclusively to the students, while students mostly divide the causes for their difficulties between their instructors and the complexity of mathematics itself (the lack of tools playing only a minor role in the survey).

Gueudet stresses the fact that the way mathematics is taught is a relevant factor in itself, coming to the conclusion that the "difficulties of the students are not individual difficulties, but consequences of [...] institutional practices" (Gueudet, 2008, p. 247)[10]. In this, she would probably agree with Tall's statement that it is "no longer viable, if indeed it ever was, to lay the burden of failure of our students on their supposed stupidity, when now the reasons behind their difficulties may be seen to be in part to be due to the epistemological nature of mathematics and in part to misconceptions by mathematicians of how students learn" (Tall, 1991b, p. 251f.). These positions stress the responsibility of educational institutions, of teachers, lecturers and researchers in education, to create learning environments and to foster learning techniques and conditions that support learners of mathematics and show understanding for their difficulties. This categorisation suits the purpose of a project drawing on *Design Research*, as it implies interventions[11].

---

[10] Here Gueudet refers to the practise of promoting only one technique to solve narrow types of tasks.

[11] In fact, (de Guzmán et al., 1998) suggest various measures to counter the difficulties observed, among them a contact desk where students can get answers for their mathematical questions.

Social aspects

As stated above, interaction is the key to building mathematical constructs; and interaction can only take place in a social context. Although the interaction between the teacher / instructor and the learner is the focus of various educational research, from the sheer quantity, there is much more interaction between individual learners – particularly at the university level, where the ratio between lecturer and students is much smaller than between teacher and pupils in a school environment. Social aspects must be considered in detail, both in their potential for creating understanding while discussing and explaining concepts, and in their ability to help learners' motivation and perseverance in a peer environment.

Yackel and Cobb (1996) have researched the sociomathematical mechanisms behind the process of becoming autonomous in mathematics in detail. They argue that sociomathematical norms play an important role and that "the teacher's role as a representative of the mathematical community" (p. 458) is central: In a classroom situation, explanations and argumentations based on the generally accepted norms are the key to building mathematical knowledge and understanding. The teacher as the authority that decides whether an argument holds or not is imperative for keeping the learning processes on track. This perspective stresses the importance of interaction between teacher and students.

At the university level, many students regularly meet in groups to work on tasks, often outside of organised learning arrangements. These group sessions can be very fruitful, as von Glasersfeld (1991, p. 176) explains, elaborating on social instruction.

> I take it this refers to "group learning", and there is a lot to be said about this.
>
> (1) Students who work at a problem together with other students have to verbalize how they see the problem and what they want to do about it. This is one way of generating reflection, which requires awareness of what one is thinking and doing. This, in turn, provides occasions for active abstraction (repeating, writing down, and learning by heart what a teacher says, does none of this).
>
> (2) Explaining something to a peer usually leads to seeing things more clearly and often to spotting inconsistencies in one's own thoughts.

And when a small group explains its "solution" (irrespective of whether it happens to be viable or not) to the whole class, this provides a wonderful opportunity for learning [...].

(3) Knowing that those you work with have no ready-made answer increases everyone's courage to try and find one.

(4) If one of the group finds an answer, this more often than not generates motivation to try a new problem.

(5) To have an inconsistency or "error" explained by a peer is far less painful than have the teacher tell you that you are wrong. Etc., etc.

To use the social aspects of learning therefore seems a sensible thing to do. As normally social interactions between students take place in an unorganised way, there is the possibility to make them more purposeful.

And there is another, wider, perspective to the social aspects of learning mathematics. It is not restricted to the relationship and interactions observable in classroom interactions, but incorporates aspects regarding society and the positions of power therein. Skovsmose, Valero, and Christensen (2009) and Halai (2014) regard mathematical knowledge as a way to gain empowerment in a political sense – as power and influence are often connected to technological and therefore mathematical skills. Alrø, Ravn, Valero, and Skovsmose (2010, p. 15) summarise that the "use of knowledge is to be seen as political, i.e., mathematical knowledge can be used in order to influence society". Although this aspect is not central to our project and research focus, it is not unrelated: As engineering is a career often aspired to by descendants from non-academic families (Bathke, Schreiber, & Sommer, 2000; Becker, Haunberger, & Schubert, 2010)[12], supporting first-year engineering students in mathematics truly contributes to their participation in society.

Affective aspects

Affect has gained a growing role in the research on mathematics education in recent years. This is mirrored in the many publications and activities on this topic, for an early overview see McLeod (1992), for more recent developments

---

[12] The total image is diverse, though, as there are different ways to gain entrance to university, various types of higher education institutions, and several kinds of engineering courses, see Schindler (2012).

Hannula, Evans, Philippou, and Zan (2004), Leder, Pehkonen, and Törner (2002), Leder and Grootenboer (2005), Roesken and Casper (2011), and Zan, Brown, Evans, and Hannula (2006). In particular, the theory of dual processes in cognitive psychology has been adapted to mathematics education, and the role of affective variables has been pointed out in this context (e.g. Evans, 2007). These perspectives provide novel views on learning processes and have done much to reach a deeper understanding of the obstacles involved. It is now generally accepted that "affect plays a significant role in mathematics learning and instruction" (McLeod, 1992, p. 575), in a more pronounced way than in other subjects, i.e. learning is considerably influenced by the affective aspects accompanying the learning process. Goldin (2002, p. 60) even ventures to say that "When individuals are doing mathematics, the affective system is not merely auxiliary to cognition - it is central". Liston and O'Donoghue (2008, p. 10) found that "Enjoyment of maths and mathematical self-concept, were the strongest affective predictors of exam results together with the level of maths previously studied". On the other hand, negative affective factors (Nardi & Steward, 2002) are feared to be omnipotent at least for less able students (Larcombe, 1985). When these are combined with deficient mathematical skills, which are steadily on the increase, as some researchers[13] report, the impact on university education can be severe.

In this context, the term affect (McLeod, 1992; Lester, Garofalo, & Kroll, 1989) indicates a broad range of phenomena, and can be differentiated into beliefs, attitudes and emotions (here ordered by decreasing stability over time, and by decrease in cognition), see Table 2.5[14], in which McLeod summarises constructs by Mandler (1984) and Snow and Farr (1987). All three major categories impact on learning outcomes and are therefore worth looking at. The next paragraph briefly summarises descriptions of the three main categories of affect.

The first category, beliefs, is further differentiated into beliefs about mathematics, beliefs about self, beliefs about mathematics teaching, and beliefs about the social context (McLeod, 1992; Mandler, 1984; Snow & Farr, 1987; Roesken, Hannula, & Pehkonen, 2011). These facets influence the way learners approach a task, the tools they employ to solve it, the perseverance they

---

[13] For example Berger and Schwenk (2006), G. Henn and Polaczek (2007), or Knospe (2013).
[14] Later, DeBellis and Goldin (1997) added a fourth category, values.

Table 2.5: The Affective Domain in Mathematics Education,
(McLeod, 1992, p. 578)

Category	Examples
Beliefs	
About Mathematics	Mathematics is based on rules
About self	I am able to solve problems
About mathematics teaching	Teaching is telling
About the social context	Learning is competitive
Attitudes	Dislike of geometric proof
	Enjoyment of problem-solving
	Preference for discovery learning
Emotions	Joy (or frustration) in solving
	non-routine problems
	Aesthetic responses to mathematics

show while doing so, the expectations they confront their teaches or lecturers with, and the meaning they perceive behind it. Beliefs are largely cognitive and tend to be stable over long periods of time – but are changeable in principle. Attitudes are less stable over time and less cognitive than beliefs (Hannula, Evans, et al., 2004). They may change more frequently and more easily. Emotions are the most unstable and less cognitive construct of the three. They may differ from situation to situation, and often for (at first sight) incomprehensible reasons. What these affective aspects have in common is that they influence achievement (Ma & Kishor, 1997; Juter, 2005; Kloosterman, 1988) via metacognition or social aspects (among others).

Mathematical beliefs offer a depth of views on the subject central to our studies. Students in their first year at university are on the one hand experienced pupils, they have passed trough more than a dozen years of mathematics instruction at school. On the other hand, they have hardly any experience in certain areas of mathematics – and thereby, certain beliefs about mathematics might be limited. Goldin (2002) gives an overview to illustrate the width of types of mathematical beliefs.

- Beliefs about the physical world, and about the correspondence of
  mathematics to the physical world (e.g., number, measurement);

- Specific beliefs, including misconceptions, about mathematical facts, rules, equations, theorems, etc. (e.g., the law of exponents, the quadratic formula, the idea that "multiplication always makes larger");
- Beliefs about mathematical validity, or how mathematical truths are established;
- Beliefs about effective mathematical reasoning methods and strategies or heuristics;
- Beliefs about the nature of mathematics, including the foundations, metaphysics, or philosophy of mathematics;
- Beliefs about mathematics as a social phenomenon;
- Beliefs about aesthetics, beauty, meaningfulness, or power in mathematics;
- Beliefs about individual people who do mathematics, or famous mathematicians, their traits and characteristics;
- Beliefs about mathematical ability, how it manifests itself or can be assessed;
- Beliefs about the learning of mathematics, the teaching of mathematics, and the psychology of doing mathematics;
- Beliefs about oneself in relation to mathematics, including one's ability, emotions, history, integrity, motivations, self-concept, stature in the eyes of others, etc.

(Goldin, 2002, p. 67f.)

Especially the last three types of beliefs are relevant when encountering university mathematics. They refer both to mathematics and to the self, thus bridging the eventual gap between what learners aspire after and what they perceive themselves to be able to. Creating a balance would go a long way to help first-year students on their way to become successful graduates.

There are other psychological concepts that have an influence on the learning of mathematics but do not necessarily come under the heading of affect, e.g. confidence, self-concept, self-efficacy, mathematics anxiety, causal attributions, effort and ability attributions, learned helplessness, and motivation, which McLeod (1992) terms "mini-theories about parts of the affective domain" (p. 583).

It would mean going beyond the scope of this chapter to elaborate on all these concepts in detail. It is obvious, though, that the above-mentioned categories of affective domains (see Table 2.5) can be applied. For example, confidence, self-concept, and self-efficacy can be classified as beliefs about self, mathematics anxiety as an emotion, and attributions can be classified as attitudes. Still, taking a closer look at these concepts yields more insight into affective influences on learning. The idea that confidence in oneself promotes learning, for example, has been widely tested and accepted (Schoenfeld, 1992). The attitude learners have towards mathematics helps achievement, and not vice versa, found Ma and Kishor (1997). This is closely connected to gender issues (Leder, 1995; Pehkonen, 1997; Hannula, Maijala, & Pehkonen, 2004), as boys tend to show more confidence and a more positive attitude towards mathematics. The statement that there is a "tendency to believe that learning mathematics is a question more of ability than effort" (McLeod, 1992, p. 575) can be understood in this context that success or failure in mathematics is more often attributed to stable (uncontrollable) than to unstable (controllable, thus influenceable) issues. This impacts on teaching concepts which postulate that success is possible, i.e. influenceable.

For our purposes, beliefs about self, which cover a range of views individuals have on how they will perform and for what reasons, are of special interest. This includes motivation (for more detail on motivation, see next section) and attribution theories, which are associated with Weiner (e.g. 1994) and have been widely researched (e.g. Bempechat, Nakkula, Wu, & Ginsburg, 1996), gender differences included (Forgasz, 1995; Hyde, Fennema, Ryan, Frost, & Hopp, 1990). The reasons an individual specifies for success or failure come in three dimensions. They can be internal or external, stable or temporary, and controllable or uncontrollable (this last dimension was added later), see Table 2.6. For example, if a student fails a mathematics test, he or she can attribute it to an external, stable cause: the task was too difficult. Or he or she can attribute failure to an internal unstable cause: I did not put enough effort into it. Attribution typically varies if applied to success or failure, and gender-specific attributions were often observed, albeit with some variation, depending on other factors like achievement or cultural background (Freislich & Bowen-James, 2001; Fennema & Leder, 1990). Males tend to attribute failure to external stable causes, and success to internal unstable causes. Females tend to do the opposite, resulting in a negative self-image, as success is attributed

Table 2.6: Attribution Model, according to Weiner (1994),
         columns = locus of causality, rows = stability dimension

	Internal	External
Stable	Ability	Task difficulty
Temporary	Effort	Luck

externally. The third dimension of controllability is an important addition, as it potentially assigns the learner an active part in the learning process.

Having explored literature on cognitive and non-cognitive theories with application to the transition to tertiary mathematics, the choice by Liston and O'Donoghue (2008)[15] was first considered as a starting point for our project for supporting first-year engineering students in mathematics. To keep the concept manageable, three affective aspects were addressed as a basis for the conceptualisation:

- beliefs (about oneself, about mathematical ability, and about the learning of mathematics, cf. the list by Goldin on page 32),

- attitude towards mathematics, and

- approaches to learning.

The last, approaches to learning, seemed the most easily accessible, it is therefore hypothesised that it may have the potential to influence the other, more hidden variables, beliefs and attitude towards mathematics. The supposition is that this will in turn influence mathematical achievement.

Self-regulation and motivation in stressful situations

For many students in a university course including mathematics, the first months are a potentially stressful time. They encounter new people, novel situations and unique challenges. Seminal research on stress has been conducted by Bandura (1977), who explored the influence of performance accomplishments,

---

[15] Liston and O'Donoghue (2008) identify five affective factors to "impact strongly on the transition to university mathematics": attitude, beliefs, mathematics self-concept, one's conception of mathematics, and approaches to learning (p. 2).

verbal persuasion, and emotions. As learning mathematics at tertiary level may not always run smoothly, the way an individual emotionally copes in perceived crisis situations is relevant, too. This has been researched in depth by (Lazarus, 1991), see Figure 2.5. Transferred to the learning of mathematics, a learner would encounter a potentially stressful environment, e.g. a task in an important examination. He or she would then primarily appraise the situation, as either positive (he or she can solve the task and interprets it as a success), dangerous (he or she anticipates failure and due consequences), or irrelevant (he or she does not care about the outcome, maybe because the mark does not count). Depending on a secondary, more rational, appraisal, in which the learner evaluates his or her resources (in reference to learning mathematics, his or her prior knowledge, heuristic skills, help available etc.), according to Lazarus (1991), stress does or does not develop. The coping (in whatever way) can focus on the problem (tackle the task, increase resources etc.), or on the emotion (avoid feeling of failure by external attributing). This results in a reappraisal of the situation which can be labelled learning - although maybe not in the desired sense of increasing mathematical skills. Beside the details of Lazarus' theory, the implications for university mathematics are clear: Learners of mathematics at university will be more successful if they have adequate resources available, and if they focus on the problem. Having lived through (and coped with) stressful situations can furthermore help to cope with future challenges. This is in keeping with theories on self-regulation and their impact on the learning process (Fox & Riconscente, 2008).

Concepts of learning styles, of metacognition, of self and affect come together in the theory of self-regulation (Vohs & Baumeister, 2011; Boekaerts, 1999; B. Schmitz & Wiese, 2006; Zimmerman, 2000; Winne, 1996; Livingston, 2003). Self-regulation is a concept that has come into focus in recent years because in a technology-based world it is becoming more and more important to adapt to new issues and to learn in all stages of one's educational and professional life, often without direct or individual support (cf. Köller & Schiefele, 2003). For an individual, to be able to regulate his or her own learning behaviour according to the needs of the situation, is a precondition for personal and social success. What is more, "being able to regulate your own learning is viewed by educational psychologists and policy makers alike as the key to successful learning in school and beyond" (Boekaerts, 1999, p. 446), and thus as the lever by which to open educational and social possibilities for everyone. The

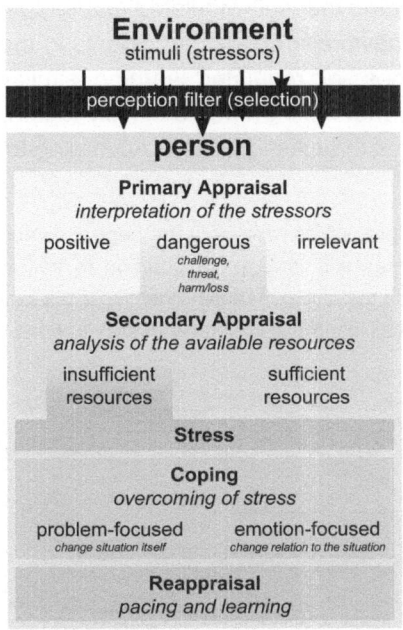

Figure 2.5: Transactional Model of Stress and Coping, according to Lazarus (1991), graphic by Philipp Guttmann, https://commons.wikimedia.org/wiki/File:Transactional_Model_of_ Stress_and_Coping_-_Richard_Lazarus.svg, licensed under the Creative Commons Attribution-Share Alike 4.0 International license.

connection between mastering self-regulation and academic success has been observed many times, e.g. by Pintrich and de Groot (1990). Findings reveal that students' cognitive reflection, as a metacognitive variable, their beliefs about mathematics, and their self-efficacy, are all correlated positively and significantly with mathematical achievement (Gómez-Chacón, García-Madruga, Vila, Elosúa, & Rodríguez, 2014). There is also evidence that metacognition impacts positively on learning strategies which in turn influences achievement (Glasmachers, Griese, Kallweit, & Roesken, 2011).

Definitions of self-regulated learning behaviour differ, though, due to the background of the individual researcher. Zimmerman (1990) describes the common understanding that self-regulated learners are "metacognitively, motivationally, and behaviorally active participants in their own learning" (p. 4). Winne (1996) outlines self-regulated learning, similarly, "as metacognitively governed behavior wherein learners adaptively regulate their use of cognitive tactics and strategies in tasks" (p. 327). According to Hannula, Evans, et al. (2004, p. 10),

> Self-regulation processes represent the central combining feature of self-system processes with affect. In addition to self-appraisals and self-judgments, these metalevel mental processes involve students' self-directive constructions, self-control and self-regulatory actions.

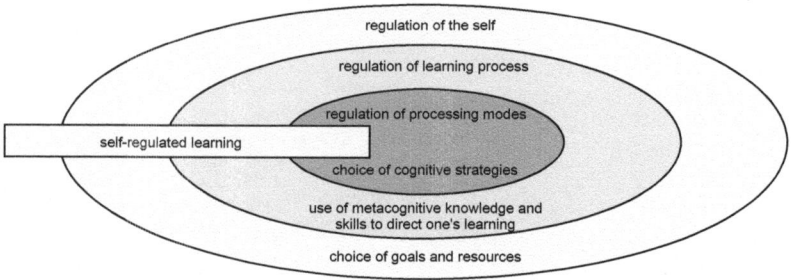

Figure 2.6: The Three-Layered Model of Self-Regulated Learning
(Boekaerts, 1999, p. 449)

A broader and more comprehensive perspective is taken by Boekaerts et al. (Boekaerts, 1999, 1997; Boekaerts, Pintrich, & Zeidner, 2000) who describe self-regulated learning in a three-layered model, see Figure 2.6. They take three schools of thought into account, referring to "(1) research on learning styles, (2) research on metacognition and regulation styles, and (3) theories of the self, including goal-directed behavior" (Boekaerts, 1999, p. 445). Each research focus is represented by a layer in the model, which allows for concentrating on single aspects as well as exploring interrelations and the big picture. The hierarchy implied by introducing layers is justified by ordering the aspects of self-regulated learning from concrete to abstract, from inner to outer layer. The inner layer refers to processing modes, i.e. choices of cognitive strategies. Examplified, processing new information would mean underlining important passages in a text, summarising them, finding examples, or connecting them to previously learned facts. Boekaerts stresses the fact that a learner has to be aware (in this context: has to know) that there are alternatives before actively making a choice. The middle layer in Boekarts' model is titled *regulation of the learning process*, meaning "use of metacognitive knowledge and skills to direct one's learning" (Boekaerts, 1999, Figure 1, here reproduced in Figure 2.6). More abstract than the first layer, this layer relates to thinking about the learning process, and comparing it to other, alternative learning processes, with the aim of increasing efficiency. Operationalised, the skills needed here are orienting, planning, executing, monitoring, and correcting (Brown, 1987; Weinstein & Mayer, 1986; Wild, 1994).

The fact that there is yet another layer in the model, can be seen in connection to the psychological origin of the authors. They are aware of the fact that not only knowledge about the learning process is needed in order to successfully accomplish learning (particularly complicated academic material), but that regulation of the self is guided by "the students' involvement in and commitment to self-chosen goals" (Boekaerts, 1999, p. 451), and by the resources they have available.

This leads us to theories on motivation and self-determination to understand what prompts people to do something. According to Deci and Ryan (1990), an individual's motivation is guided by the expectation if his / her basic psychological needs will be met. These comprise competence (the need to experience control over what is regarded as important), autonomy (the need to perceive oneself as actively making decisions), and relatedness (the need to feel connected to others). The authors (Deci & Ryan, 1990; Ryan & Deci, 2000; Deci & Ryan, 2000) also elaborate on different forms of motivation (from amotivation over four different grades of extrinsic motivation to intrinsic motivation, see Figure 2.7) which they connect to the degree to which these basic needs are fulfilled. Ideally, all three needs are satisfied to an individually acceptable degree, which leads to volition (expressly wanting to reach the learning aims), motivation (feeling able and eager to reach the learning aims) and engagement (in actions that lead to reaching the learning aims), see Deci and Ryan (1990). The presence of these attitudes then results in increased effort and performance. If an individual does not expect that participation in an activity will fulfil his or her basic psychological needs, he or she might probably not even endeavour to engage in the activity. The awareness of these circumstances has even found its way into human resource management and helps to motivate employees through catering for their need to experience social support and leeway for their actions (Nerdinger, 2014).

Behavior	Nonself-determined					Self-determined
Type of Motivation	Amotivation		Extrinsic Motivation			Intrinsic Motivation
Type of Regulation	Non-regulation	External Regulation	Introjected Regulation	Identified Regulation	Integrated Regulation	Intrinsic Regulation
Locus of Causality	Impersonal	External	Somewhat External	Somewhat Internal	Internal	Internal

Figure 2.7: The Self-Determination Continuum (Deci & Ryan, 2000, p. 237)

More specifically, Heckhausen has explored the mechanisms of achievement motivation (Heckhausen, 1977, 1989; Heckhausen & Heckhausen, 2010). He describes four phases: "(1) the initial *situation* as appraised, (2) the person's own *action*, (3) the *outcome* of the action or of the situation, and (4) the *consequences* of the outcome, with their various incentives determining the value of the outcome" (Heckhausen, 1977, p. 286f.). The crucial point is in what an individual anticipates will be the probable outcome of his or her action. The incentives that ideally lead to evaluating the outcome as valuable enough to perform an action are

> affective states people expect to experience after or while performing a certain behavior [...] In a learning context an incentive may be, for example, the anticipated pride (affective goal state) regarding a good performance in an exam. This anticipation is likely to lead to revising for this exam (goal-directed behavior). Depending on the positive (e.g., pride after passing the exam) or negative (e.g., disappointment after failing the exam) value of the anticipated affective state, people develop approach or avoidance tendencies towards the exam situation.
> (Schüler & Engeser, 2009, p. 340)

The distinction between an outcome and the consequences of an outcome is necessary (Heckhausen, 1977, p. 286) because an outcome (e.g. an 80% score in an exam) can have different consequences for different individuals (e.g. feelings of pride or shame, depending on the expectations), and because an outcome can bring about more than one consequence (e.g. concerning self-evaluation, other-evaluation and superordinate goals).

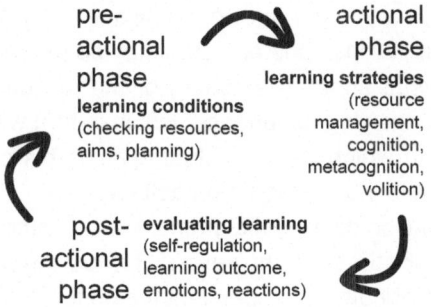

Figure 2.8: The Cycle of Self-Regulation

The consequences experienced in a learning situation can in turn influence the following learning experiences, when self-regulation is viewed more as a cyclic process, as described by B. Schmitz and colleagues (Landmann & Schmitz, 2007b; B. Schmitz, 2003; B. Schmitz & Wiese, 2006), based on the work of Zimmerman (e.g. Zimmerman, 2000, 1990), and involves "goal-setting, monitoring and regulation" (B. Schmitz, 2003, p. 221) in a three-step-cycle, see Figure 2.8.

According to Landmann and Schmitz (2007b), the process of self-regulation itself is divided into three parts: the pre-actional, actional, and post-actional phases. The authors elaborate that before a learning activity is started (during the initial appraisal of the situation, according to Heckhausen), the individual has certain resources at his or her disposal, aims at certain goals to reach with the learning activity, and plans what to do exactly, influencing how the actional phase, the immediate learning activity, is approached. As depicted in this theory (B. Schmitz & Wiese, 2006; B. Schmitz, 2003), the actional phase is characterised by the individual's strategies, meaning resources (e.g. works of reference), cognition (knowledge and skills), metacognition, and volition. This influences the post-actional phase (in Heckhausen's terms, the outcome and its consequences), where the individual assesses his or her learning, the way they can control the learning process, by the results that are reached, by emotions and the reactions to these emotions. According to this line of thinking, if the learning activity produces ample satisfying results, there will probably be positive emotions, and the individual will continue the activity until the end without big difficulty. But if the learning activity turns out to be a frustrating experience that cannot be remedied by employing available resources, emotions may falter, and the individual may give up the learning activity (Schmidt et al., 2011). One outcome may be the realisation that further help is needed (consult another textbook) or that an easier or shorter task may lead to a more satisfying experience, which in turn will influence the next pre-actional phase, in which the individual reviews his / her resources and learning aims and plans the next learning activity.

Viewing self-regulation as a (cyclic) process is no contradiction of the three-layered model described above, however. Rather, the view of the process raises the awareness of how interdependent the concept is – whereas the view of the layers accounts for its complexity. This results in the insight that success or failure of self-regulation in individuals depend on different features, among them, learning strategies, learning styles, and personality (Cervone, Mor, Orom, Shadel, & Scott, 2011; Zimmerman, 1989).

Coping with stress, maintaining motivation, and regulating the self via metacognition are the key challenges when understanding learning in academic contexts. They are multi-faceted processes and involve many variables. All these constructs bear the common feature of thinking about one's own thoughts and perceptions, i.e. of having reached Piaget's formal operations stage, see Table 2.2. To tackle the problem of improving learning mathematics at university, it is essential

- to point to resources sufficient enough to overcome the challenge (cf. Lazarus),

- to offer a choice of alternatives in organising the learning process (following Boekaerts),

- to emphasise that the decision which alternatives to employ is an autonomous one (as understood by Deci and Ryan, (1990)),

- to discuss the possible outcomes of different actions (according to Heckhausen), and

- to convey in the learner feelings of control over possible outcomes (termed competence by Deci and Ryan, (1990)).

2.1.3 Approaches to overcome transition problems

The attempts to overcome the obstacle of transition from secondary to tertiary education in mathematics vary according to the underlying hypothesis about what forms the core of the problem, the perspective of the initiators, and the means at their disposal. A number of predictors for success in university mathematics courses have been identified: Among these are the mathematics mark reached in school (Rach & Heinze, 2013; Rach, Siebert, & Heinze, 2016), mathematical competence (G. Henn & Polaczek, 2007), the use of learning activities offered by the university, the teaching quality in terms of cognitive activation or discussion of alternative solutions, deep learning strategies, and motivation for / interest in mathematics, see Griese, Glasmachers, Kallweit, and Roesken (2011); Gómez-Chacón et al. (2014), and, for an empirically backed overview, Blömeke (2016), and the meta-analysis by Trapmann, Hell, Weigand, and Schuler (2007). Other researchers are looking for reasons for failure and find early and constant use of calculators (Weinhold, 2014) or lowered demands in school curricula (G. Henn & Polaczek, 2007). This list can be complemented with the reasons students give when asked why they decided to drop out of a university mathematics course (Heublein & Barthelmes, 2010): they most often report difficulties to meet the standards (33%), or a lack of motivation (25%). In a survey among students from universities and universities of applied sciences, Bargel (2015, p. 28, Figure 6) found that the top requests for improving study conditions were more courses in small groups (2013: 25%), a stronger practical relevance (2013: 24%) and fixed tutorials (2013: 20%). In 2007, however, it was a different list. Then, bridging courses came first (39%), before more courses in small groups (36%) and a stronger practical relevance (29%). This development takes account for the fact that bridging courses have become a common feature.

It is important to note that there are many projects around aiming at overcoming transition problems in mathematics (Dunn, Lo, Mulvenon, & Sutcliffe, 2012; Hoppenbrock, Biehler, Hochmuth, & Rück, 2016)[16]. This in itself is an indication of how huge this problem is, how many people are affected, and of how urgently a solution is needed. The obstacle exists all around the world, as the fact shows that similar approaches can be found all over Germany, in Europe, and in many other countries.

---

[16] A list (under certain criteria) is compiled in section 2.1.5 later in this work.

For reasons of completeness, at first some categories of possible interventions are described that are not directly connected with support projects for first-year students. To paint the whole picture, these arrangements often complete and complement a university's or a district's catalogue of interventions to help young people on their way to participate in society, according to their capability and inclination.

To start out, it seems advisable to provide more information on university courses at school level. There are initiatives that aim at improving the information about university courses involving mathematics that are available to students[17]. Thus, their prospects are expected to be more realistic. Despite an ambitious school curriculum, it is hard to impart the concept of mathematics as a complex science at high school level. Nevertheless, multifarious facts are available, online and otherwise, that describe university science and engineering courses and that aim at conveying the idea of hard work, long hours, and deep thinking. In addition to this, universities offer open days where students can partake in lectures and lab work. This first-hand experience is supposed to not only attract excellent students, but also to inform average ones.

Entrance tests for certain university courses are another suggestion. Some universities use tests as a means to choose their students. In Germany, this is not usually an option (for STEM subjects[18]), but universities can take an applicant's school mark in mathematics as a further criterion for awarding places to study (in combination with the average mark in the school-leaving examination, beside other criteria such as waiting time). In other countries, this is completely different: Depending on their status and ranking, universities can either choose from among the very best of a generation[19], or contend themselves with whoever is able to pay the college tuition fees.

---

[17] For example www.studifinder.de for North Rhine-Westphalia, which combines the search for suitable courses with tests and customised preparation courses, for Germany see also https://studieren.de, http://www.hochschulkompass.de/, http://www.studienwahl.de and http://www.wege-ins-studium.de.

[18] For medicine, there is a tradition of entrance tests in Germany.

[19] The US-American Ivy League universities are famous for accepting below 10% of applicants, with tuition fees ranging around 35,000$ a year, see https://www.studential.com/applying/studying-abroad/USA/ivy-league-universities.

Third, some universities try to reform and improve their courses, e.g. the mathematics department at Ruhr-Universität Bochum[20]. They enhance first year lectures according to didactic demands: Lecturers and teaching assistants are coached to openly show interest and respond to newcomers' problems, to visualise their explanations and give illustrating examples, to warn off typical misconceptions and to be aware of characteristic mistakes. The accompanying tutorials are also an objective for innovations – and Szczyrba and Wiemer (2011) believe they can be a motor for good teaching. It is almost standard procedure today to try out technical support for lectures in different ways: tasks can be set online, lectures are audio- or video-recorded, digital versions of scripts are uploaded, e-mail support is offered, and much more. Modern software even allows for individual and constructive feedback, a huge step forward from the true/false single or multiple choice tasks from before, like the STACK plugin for the learning platform moodle[21]. More information on technological support can be found later in this section.

Formal relief for the start at university is another approach to support average or weak first-years. Several universities have taken the fact into account that problems mostly occur at the beginning of a university course, and that, once these initial transition problems have been overcome, most students' courses indeed run rather smoothly. Logically, they try to make the start easier by allowing for (more) unsuccessful attempts in exams, by enabling students to reduce the workload in their first year, or by giving them more time to adapt to the new learning conditions. For example, Ruhr-Universität Bochum has devised a model where mathematics students concentrate on linear algebra and take calculus later, in order to take pressure from students who normally would have to attend both lectures in their first year. So far, there is no systematic evaluation if this idea has indeed raised the number of graduates.

As mentioned above, the use of modern technology offers a whole new range of support devices. It makes sense, therefore, that many of the foregoing projects make use of modern technology in the form of online support, e.g. by using learning management systems, software that allows users to upload and share files and links, with comments, tasks and assessment tools. In education,

---

[20] http://www.ruhr-uni-bochum.de/imperia/md/content/mathematik/service-zentrum/szma
_schulung.pdf
[21] https://moodle.org/plugins/qtype_stack

moodle is one of the most popular. It is free, open source, and features count-
less plug-ins for different purposes. But also in the commercial sector, there is
hardly any popular sharing site that does not serve those with mathematical
questions. For example, www.youtube.com gives you 5,600,000 hits for "math
help" (in November 2013), the top ones sporting more than 2,000,000 clicks,
while some German mathematics help videos on www.youtube.de accomplish
around 145,000 clicks. The current generation of students has grown up with
the perception that the answer to any possible question, however specific it
may be, is usually just one google search away. But even before the Internet
was available everywhere and all the time, there was software to support visualisation and thus understanding. The availability of computers in general and
dynamic geometry software and graphing software in particular have backed
this development, even though they may not have conquered classrooms as far
and wide as might be desirable. The chances offered by these devices were
discerned decades ago, as McLeod revealed when expressing the view that
"technology can play an important role in changing beliefs about mathematics
and possibly even in improving attitudes toward mathematics" (1992, p. 588).

As early as (1998), de Guzmán et al. suggested offering learning centres
or help desks to students in need of mathematical support. An intervention
like this is usually part of almost every support project (e.g. M. Schmitz &
Grünberg, 2016). Ideally, students can get their questions on current homework
assignments or the lecture answered in an informal surrounding. Often, these
help desks or learning centres are manned by more advanced students, so
that the atmosphere is more relaxed than it could ever be in a professor's office
hour. Apart from face-to-face explanations, often there are additional learning
materials available, e.g. summaries of basic domains, example exercises, a
collections of test papers from the past, explanatory videos, or Internet access.
Like many other universities (for a large-scale Irish survey on the impact of
support centres, see Ní Fhloinn, Fitzmaurice, Mac an Bhaird, & O'Sullivan,
2014), Ruhr-Universität Bochum has help desks both for service mathematics
lectures (established 2007[22]) and for mathematics majors (established 2012[23]).
Other universities have more resources and sport award-winning facilities,

---

[22] http://www.ruhr-uni-bochum.de/helpdesk-mathematik/
[23] http://www.ruhr-uni-bochum.de/ffm/Lehrstuehle/stochastik/lernzentrum.html

like Loughborough University with its Mathematics Learning Support Centre[24] (Jaworski, 2008).

Preliminary or bridging courses offered between school and university or at the beginning of a university course are among the most popular measures a university can take to support students. Many universities offer voluntary bridging courses: Before lectures start, new students are invited to attend a course, usually lasting a few weeks, where what is considered general school knowledge in mathematics is revisited, mostly in the form of a lecture, sometimes in combination with tutorials and homework, usually in groups of less than 30 students. As many as 48% of engineering students report that there were bridging courses at their university, and that they took part (Bargel, 2015, p. 38). Although this seems promising, details like how to address an appropriate target group are not always solved. Those students who enrol early enough to feel addressed by bridging courses may be above average. What is more, bridging courses require students to visit university regularly *before* their regular time there has started, and they normally do not warrant any credit points. Both reasons result in not reaching the target group the bridging course was initially intended for, namely average (or even below average) students with an incomplete mastery of school mathematics. Although there is a unique consensus that first-year students of mathematics or of subjects related to mathematics often lack basic skills in high school mathematics, such as rearranging terms, solving equations, fractions, powers, and roots, the efficacy of bridging courses has not yet been universally observed, but positive effects are there (Heiss & Embacher, 2016; Greefrath & Hoever, 2016). For example, Kürten, Greefrath, Harth, and Pott-Langemeyer (2014) found significant differences between students who had attended a bridging course and those who had not only in selected subgroups of their sample. Despite the statistics, students attending bridging courses often report to feel their time well-spent; according to Bargel (2015, p. 48) the large majority judge bridging courses as useful (65%) or partly useful (31%).

The idea to specifically review the skills and concepts that have not been mastered (e.g. modes of thought, see Hoffkamp, Paravicini, & Schnieder, 2016), and at the time when they are needed, expands the notion of bridging courses into courses accompanying students during their first months at university. This

---

[24] http://www.lboro.ac.uk/departments/mlsc/

kind of course can also incorporate some of the ideas described above, like the use of learning platforms or other technology (like Ellis, Goodyear, Rafael, & Prosser, 2008, who explore approaches to learning and understanding in a blended-learning environment), as well as reviewed didactical concepts with purposely trained and carefully chosen staff. It needs thoughtful planning, though, as it is meant to complement and not to replace the regular courses at university.

Table 2.7: Description Categories for Objectives, according to WiGeMath, translation by author

Learning objective
Knowledge-related objectives
Basic mathematical knowledge and skills
Academic mathematical knowledge and skills
Terminology
Action-related objectives
Mathematical methods
Learning strategies
Learning behaviour (e.g. study time)
Attitude-related objectives
Beliefs
Affective features
Practical relevance
Mathematical enculturation
System-related objectives
Reviewing skills and abilities
Introducing academic methods and procedures
Imparting (mathematic-related) learning strategies
Quality of objective
Clarity resp. concreteness of objectives
Transparency and propagation of objectives

A conceptualisation to describe a framework for the reconstruction of supportive interventions is currently being developed in the course of the

WiGeMath project[25] at Leibniz Universität Hannover and Universität Paderborn, which researches conditions of effect and success of support interventions for mathematics-related learning (*Wirkung und Gelingensbedingungen von Unterstützungsmaßnahmen für mathematikbezogenes Lernen in der Studieneingangsphase*). According to presentations at the kickoff workshop in autumn 2015 and at a conference in spring 2016 (Colberg et al., 08.03.2016), a first version of the framework categorises support interventions in regard to their objectives, the interventions and their characteristics, and general parameters. The categorisation of the objectives is presented in Table 2.7.

The second group of categories, which refers to the interventions and their characteristics, comprises structural characteristics (format, duration, temporal structure and appointed times), didactical elements (didactic principles and guidelines, exercises, interactive and social forms, and summative vs. formative tests), and characteristics of the teachers or teaching teams (number, status, role, as well as subject-specific, subject-didactic or university-didactic qualification), see Colberg et al. (08.03.2016). The last group of conditions covers general parameters, like number of students and their characteristics (age, gender etc.), mathematical foreknowledge, genesis and development of the intervention, curricular embedding (optional or mandatory), organisational characteristics (e.g. selective measures), rooms (e.g. flexibility of furniture), financial conditions, and characteristics of the teaching and learning culture (epistemological aspects of the subject-specific culture, interaction between teachers / lecturers and students, students' and lecturers' expectations). The WiGeMath framework is still in the stage of being tested and will be adapted to improve the fit to the interventions that are explored in detail. At this point, however, the impression is that all bases are covered, and that this framework will serve not only to categorise support projects, but also to enable evaluating their impact and success, as planned. For the research purposes at hand, the framework provides a reference frame to classify the interventions in the project MP[2]-Math/Plus.

Furthermore, the possible factors influencing effect and success of support projects listed in the WiGeMath framework can serve to check our hypothesis for completeness and detail. WiGeMath have identified the influence factors as presented in Table 2.8, with the main categories person-related

---

[25] http://www.hochschulforschung-bmbf.de/de/1901.php

Table 2.8: Possible Impact and Effect Variables for the Evaluation of
Mathematics-Related Support Interventions, according to WiGeMath,
translation by author

Person-related factors
(Socio)demographic objectives
Cognitive / Metacognitive characteristics
Motivational characteristics
Environment-related factors
General conditions
Structural characteristics
Teaching and Learning activity
Learning process
Didactical elements
Teacher / Lecturer
Impact variables
Cognitive impact
Motivational and affective impact

factors, environment-related factors, teaching and learning activity, and impact variables.

All these projects and initiatives aim at bringing about change in the form of a better kind of learning in mathematics: be prompter, offer more specific help, make the learners more content and more motivated, or help them understand the concepts on a deeper level. The research project at hand will be classified according to the categories and framework presented in this section.

2.1.4 Specifities of engineering mathematics

The Mathematics Working Group of the European Society for Engineering Education (SEFI) has the aim to foster exchange and to compile documents offering orientation (Alpers, 2016). Their consensus on competencies, curricula, and assessment is a qualified basis for mathematical education for aspiring engineers. SEFI's understanding of mathematical competence is based on Niss (2003), who states

Mathematical competence then means the ability to understand, judge, do, and use mathematics in a variety of intra- and extramathematical contexts and situations in which mathematics plays or could play a role. Necessary, but certainly not sufficient, prerequisites for mathematical competence are lots of factual knowledge and technical skills [...] (p. 6f.).

A. Entwistle and Entwistle describe the importance and implications of understanding mathematics

not as a cognitive process, but as an experience. It involves a feeling of satisfaction as sets of information and ideas are brought together into a coherent whole. It also creates a feeling of confidence that the understanding reached can be used to construct explanations or to solve problems in novel contexts. The development of understanding may also have a social component through the discussions which lead to the negotiation of shared meaning (p. 18).

This justifies the central position understanding has in the learning of mathematics. After a first version of a curriculum document for engineering mathematics from 1992, which according to Alpers (2016) was little more than a list of the subject-specific content[26], the more recent framework document (Alpers et al., 2013) has accomplished the step to a competence-based description of learning outcomes. These competencies comprise thinking mathematically, reasoning mathematically, posing and solving mathematical problems, modelling mathematically, representing mathematical entities, handling mathematical symbols and formalism, communicating in, with, and about mathematics, and making use of aids and tools[27] (Alpers et al., 2013, p. 13f.) – meant to overlap. In conformity with Niss' above statement, the SEFI framework (Alpers et al., 2013) contains a list of subject-specific factual knowledge and skills as well, ordered from core level 0 resp. 1 to level 2 and 3. The level descriptions include:

Core Zero [...] comprises such essential material that only minor omissions are acceptable. [...] Core level 1 comprises the knowledge and skills which are necessary in order to underpin the general Engineering Science that is assumed to be essential for most engineering graduates. [...]

---

[26] „Das Curriculum-Dokument [...] bestand zum größten Teil aus aus einer Liste von zu behandelnden Themen" (Alpers, 2016, p. 645).

[27] The analogy to the competencies in the German competence model of education standards (see section 2.1.1) stems from Niss' participation in the PISA studies.

> Level 2 comprises specialist or advanced knowledge and skills which are considered essential for individual engineering disciplines. [...] Level 3 comprises highly specialist knowledge and skills which are associated with advanced levels of study and incorporates synoptic mathematical theory and its integration with real-life engineering examples (p. 20f.).

It is notable, though, that even core level 0 (which covers more than six pages listing learning objectives completing the phrase "As a result of learning this material you should be able to", p. 23–29) has a considerable number of items starting with "understand" (e.g. "understand the role of the arbitrary constant" in the section on indefinite integration, p. 26) and consistently addresses the learner, and not the teacher / lecturer. This highlights the change to describing learning outcome (and not input) and the personal responsibility of the learner, true to Alpers' belief that "If we cannot communicate to the learner what we expect them to learn, we cannot demand or expect that they learn it" (Alpers et al., 2013, p. 69).

In a similar frame of mind, the advisory board of teachING-learnING.eu has formulated eight theses on next generation engineering[28] that back this position. Particularly the first three theses draw a progressive picture of the future of engineering education:

> Next generation engineering education will prepare graduates to solve open problems and well as competing requirements. This is based on mastery of problem definition and analysis skills, domain specific knowledge in required areas, interaction collaboration, and self management skills as well as self guided initiative and creativity. [...]
> Next generation engineering education will be aware of the fact, that especially the first year of study is crucial for the personal development and thus requires personal coaching of the students by experienced mentors [...].
> Next generation engineering education will move from teacher centered (passive) to student centered (active) learning. [...] [29]

The question of how to assess these kinds of competencies arises naturally, as traditional written examinations do not seem appropriate. Students tend to

---

[28] http://www.teaching-learning.eu/fileadmin/documents/News/Theses_AdvBoard_2011_news .pdf
[29] http://www.teaching-learning.eu/fileadmin/documents/News/Theses_AdvBoard_2011_news .pdf

think and work with a strong orientation towards the examination, "What you get is what you assess" (Alpers, 2016, p. 651). In his *Mathematics Curriculum for a Practice-oriented Study Course in Mechanical Engineering*, Alpers (2014) gives concrete examples of how assessment can follow a competence-based curriculum, for example by splitting assessment into a classical written part assessing the "capability of performing short mathematical reasoning along lines encountered before or performing standard problem solving routines" (Alpers, 2014, p. 64), and a second part including "larger assignments, project work documentations and oral presentations" (p. 65)[30]. The SEFI framework finds reasons why a student should only pass if he or she has mastered every task from the first part, neglecting "minor 'numerical' errors" (Alpers et al., 2013, p. 70). Other research is dedicated to "Conceptions of Understanding in Engineering Mathematics" (Khiat, 2010), to exploring theoretical frameworks for mathematical competencies for engineering students (Hochmuth, Roesken-Winter, & Jaworski, 2013; Noss & Kent, 2002), or to developing measurement instruments for them (Neumann et al., 2015). In more or less complex applications, future engineers are expected to use mathematical models and techniques to solve open practical problems which can be described not only by the number of competencies involved, but also by the degree of coverage ("the extent to which the person masters the characteristic aspects"), the radius of action ("contexts and situations in which a person can activate" a competency), and the technical level (indicating "how conceptually and technically advanced the entities and tools are with which the person can activate the competence") needed (Niss, 2003, p. 10).

In contradiction to this mode of thought, Dreyfus (1991) finds that "what most students learn in their mathematics courses is, to carry out a large number of standardized procedures, cast in precisely defied [sic (defined)] formalisms, for obtaining answers to clearly delimited classes of exercise questions. They thus acquire the capability to perform, albeit much slower, the kind of operation which a computer can perform" (p. 28). This hints at the perception that there is a tendency to focus strongly on procedure, be it because procedural tasks are easier to grade, or because there is a general consensus that procedural knowledge is at the basis of mathematical competence. In contrast to students

---

[30] Alpers et al. (2013) are well aware of the problem of resources, however.

majoring in mathematics, engineering students do not have to conduct formal proofs, but to calculate demands on material properties, statics, and solidity.

But, as demonstrated in section 2.1.2, the universally accepted way of learning mathematics is to (re)construct each piece of knowledge by studying objects in order to abstract their properties, by building concepts, and by logical deductions from these concepts (Tall, 2004). It also involves discovering discrepancies and accommodating previous mental concepts accordingly (Piaget, 1973), which implies bringing discrepancies to light and discussing them. This requires active participation of the learner, guided and challenged by a teacher or lecturer (Tall, 1991b). In keeping with this, engineering curricula contain much more than just technical and procedural requirements (albeit no formal proofs), as the examples from Neumann et al. (2015) and Alpers et al. (2013) show. The question remains if university mathematics courses focus on the entirety of competences, or on what is sometimes perceived as the foundation of mathematics, namely basic skills and techniques. This particularly applies to the support projects for first-years, which will be elaborated upon in the next section.

2.1.5 Projects at other universities

As presented on in section 2.1.3, there are many ways in which to deal with the problem that students tend to experience difficulties in mathematics when starting a course at university. These considerations have led a huge number of universities to initiate various projects in order to overcome the obstacle. To attempt to compile a comprehensive list of who does what in this field would be presumptuous, so the elaborations in this section are restricted under the following criteria.

- The initiative caters for mathematics for (not necessarily exclusively) engineering students.
- It concentrates on the first semester or first year at university.
- It exchanges and discusses ideas in an organised way with other initiatives working in the same field.
- The intervention is connected with research in mathematics education.

– It is located at a German university.

This implies that there is still variety in the kind of university (private or state university, university of applied sciences, technical college etc.), in the specific engineering course (e.g. machine engineering, civil engineering, environmental engineering, industrial engineering), and in the particular focus of the initiative (e.g. target groups, size of project, specific learning activity).

According to their homepage[31], the competence centre for university didactics in mathematics (*Kompetenzzentrum für Hochschuldidaktik Mathematik*), khdm, is a common institution of the universities at Kassel, Lüneburg and Paderborn, and has been active since 2010. Their central fields of work include

– investigating the teaching and learning of mathematical modes of thought and practise,

– investigating students' attitudes and learning behaviour,

– developing and investigating innovations from the areas of teaching and learning methods and digital media, and

– revising and innovating existing university curricula in relation to content under the aspect of competence and recipient orientation.

khdm has a working group on engineering mathematics (AG Ing-Math[32]), concentrating on mathematics for first-year engineering students. Their central projects[33] deal with modelling in machine engineering and with situational acquisition of mathematical knowledge in electrical engineering. Other associated projects research requirements in mathematical problem-solving in electrical engineering (part of the bigger project KoM@ING[34]), produce mathematical video lessons for students of electrical engineering (LEMMA[35]), and develop multimedia material for virtual preliminary courses for STEM subjects (VEMINT[36]).

---

[31] www.khdm.de/

[32] www.khdm.de/ag-ing-math/

[33] https://www.khdm.de/fileadmin/khdm/Kolloquien_und_Oberseminare/Poster/Ing_110117_A4
.pdf

[34] www.kom-at-ing.de/

[35] https://www.khdm.de/ag-ing-math/lemma-lehrinnovationen-zur-mathematikausbildung-in-der
-elektrotechnik/

[36] www.vemint.de

In connection and collaboration with khdm, the above-mentioned project WiGeMath was founded. It aims at a conceptualisation to describe a framework for the reconstruction of supportive interventions. To reach this aim, they have brought together several representatives active in support projects in mathematics. In order to test the framework, some of the projects will be chosen to collect more data. The result is expected to deliver a theoretical framework and insights into what is relevant for support projects to be effective and successful. Thus, WiGeMath does not itself design support projects, but takes a general perspective. The WiGeMath project partners, supported by funds from the Qualitätspakt Lehre[37], were chosen from among the several initiatives contacted, many partaking in the kickoff meeting in 2015.

– RWTH Aachen, with 42,000 students the biggest university for technological courses in Germany: RWTH Aachen has offered preliminary courses in mathematics to their students for decades and has accumulated expertise and experience in teaching first-year STEM students in mathematics.

– MINT-Kolleg Baden-Württemberg[38], a joint college of KIT (Karlsruher Institut für Technologie) and Universität Stuttgart: The MINT-Kolleg is dedicated to improving the subject-specific preconditions in the transition from school to university. It offers preliminary courses, a help desk, information for those interested in studying a STEM subject, and tests.

– Ruhr-Universität Bochum: With more than 14,000 students attending courses in STEM subjects (about half of them enrolled in engineering courses[39]), the department of mathematics at RUB offers preliminary courses, a special help desk for service mathematics, and the support project MP^2-Math/Plus/Practice.

– Technische Universität Darmstadt: Its 26,000 students[40] mainly study STEM subjects; it offers individual information and activities for prospective students as well as bridging courses. There is research of the

---

[37] http://www.qualitaetspakt-lehre.de/
[38] http://www.mint-kolleg.de/
[39] http://www.ruhr-uni-bochum.de/universitaet/fakten/menschlich/index.html
[40] http://www.tu-darmstadt.de/universitaet/selbstverstaendnis/zahlenundfakten/index.de.jsp

transition from secondary to tertiary education in mathematics, particularly diagnostic tests (Schaub & Bruder, 2015).

– Technische Universität Dortmund: The majority of its around 30,000 students attend courses in STEM subjects[41]. TU Dortmund offers three different preliminary courses in mathematics[42] which especially aim (a) at students of physics, computer science (and others), (b) at students majoring in mathematics, and (c) at students of engineering and chemistry (and others). TU Dortmund also tests didactical concepts for motivating students who attend service lectures in mathematics.

– Universität Hamburg: Students can choose between more than 150 different courses (though none in engineering) and has over 40,000 students (more than 13,000 in science and mathematics[43]). There is one preliminary mathematics course that STEM students are advised to attend, including an online assessment; both are optional. The research groups for didactics are part of the faculty of education.

– Leibniz Universität Hannover: Of the more than 25,000 students, about a third are enrolled in STEM subjects[44], engineering courses are popular. The university is home of a research project that researches competences in mathematics-related engineering application tasks for first-year students (*Kompetenzen bei mathematikhaltigen ingenieurwissenschafltichen Anwendungsaufgaben in der Studieneingangsphase*[45]), and a project to explorate the development of interest in mathematics during the first year at university (*Interessenentwicklung im ersten Studienjahr*[46]) which are both connected to khdm.

– Universität Kassel: The university has 24,000 students, about a third are enrolled in STEM subjects. It offers a learning centre as well as prelimi-

---

[41] http://www.tu-dortmund.de/uni/Uni/Zahlen__Daten__Fakten/Statistik/Publikationen/ Studierendenstatistik/StuSta_SoSe15_web.pdf

[42] http://www.mathematik.tu-dortmund.de/sites/vorkurs-mathematik-2015

[43] www.uni-hamburg.de/beschaeftigtenportal/services/statistik/download/up-pvv-u-2012w.pdf+ &cd=2&hl=de&ct=clnk&gl=de

[44] https://www.uni-hannover.de/fileadmin/luh/content/strat_controlling/statistiken/ studierendenstatistik/studierendenstatistik_wisem_2015_2016.pdf

[45] https://www.khdm.de/ag-ing-math/diss-joerg-kortemeyer/

[46] https://www.khdm.de/ag-bagym-math/entwicklung-des-mathematikinteresses-im-ersten -studienjahr/

nary and bridging courses for first-year engineering students and houses a research project on working and learning strategies in mathematics at university (for students with a major in mathematics *Arbeitsweisen und Lernstrategien im Mathematikstudium*) in the context of khdm[47].

- Universität Paderborn: Like Universität Hannover, it is part of khdm and therefore has a share in the projects WiGeMath, VEMINT, KoM@ING, and other research. Universität Paderborn has almost 20,000 students, of which 2,800 are enrolled in machine engineering courses alone[48], in addition to courses in electrical engineering and other STEM subjects. They also offer a learning centre[49] and bridging courses.

- Universität Ulm: Of Ulm's more than 11,000 students, 8,000 are enrolled in STEM subjects[50]. There is research on alternative evaluations and examination formats, e.g. in the bachelor course *Computational Science and Engineering*. The university offers bridging courses and a training camp for prospective students[51].

- Universität Würzburg: As one of the oldest universities in Germany (first founded 1402, closed 1413, re-founded 1582), Würzburg has about 28,000 students, with a minority of a few thousand studying STEM subjects. There is research on metacognition in the learning of mathematics (Mungenast, 2015).

Two partner universities, Carl von Ossietzky Universität Oldenburg and Philipps-Universität Marburg, are not included here because they do not offer engineering courses.

Lehre[n] (http://www.lehrehochn.de/home/) is an alliance which selects a focus every year for a council (*Kolleg*). In 2013 the focus was on mathematics in engineering education. Six project teams from different universities were

---

[47] https://www.khdm.de/ag-bagym-math/diss-goeller/
[48] http://mb.uni-paderborn.de/presse/zahlen-daten-fakten/
[49] https://lama.uni-paderborn.de/en/lernzentrum.html
[50] http://www.uni-ulm.de/fileadmin/website_uni_ulm/studium/Studierendenstatistik/WS2015/
Statistik6_WS15_16.pdf
[51] http://www.uni-ulm.de/misc/unitrain.html

chosen for the council[52] and are listed here; several more were invited to a poster presentation[53].

- University of Applied Sciences Aachen with the project *Fördern und Fordern – Installation eines semesterbegleitenden Anpassungskurses*, support and demand – installation of an accompanying adaptation course,

- Technical University Berlin with the project *Tumult*, a multimedia blended-learning project for first-year engineering students,

- Ruhr-Universität Bochum with the project MP^2-Mathe/Plus/Praxis,

- University of Applied Sciences Hamburg with the project *Themenwochen zur Verknüpfung der Mathematik, Elektrotechnik und Physik im ersten und zweiten Semester*, topic weeks for the conjunction of mathematics, electrical engineering, and physics in the first and second semester,

- Ostfalia University of Applied Sciences with the project *MF&FM – Mehr Feedback und formative Assessments in der Mathematik*, and

- Technical University Vienna with the project *AKMATH/GKMATH – Auffrischungs- und Grundkurs Mathematik an der Technischen Universität Wien*, refreshment and basic course in mathematics.

The aim of Lehre^n is to create a trustworthy surrounding for a community of practice to exchange ideas and concepts. Representatives of the participating projects met several times to enable contact and collaboration.

The dghd (*Deutsche Gesellschaft für Hochschuldidaktik*, German society for university education) unites a considerable number of societies for university education, for all kinds of subjects. In particular for mathematics in engineering courses, there is teachING-learnING.eu[54], with offices in Aachen, Bochum, and Dortmund. teachING-learnING.eu describes itself as competence and service centre for teaching and learning in engineering sciences (*Kompetenz- und Dienstleistungszentrum für das Lehren und Lernen in den Ingenieurwissenschaften*) and offers a broad variety of information, meetings,

---

[52] http://www.lehrehochn.de/fileadmin/user_upload/mathing/LehreN_MathIngBroschüre2014.pdf
[53] www.hrk-nexus.de/uploads/media/Posterdokumentation.pdf
[54] http://www.teaching-learning.eu

conferences, workshops and working sessions for all groups involved in engineering, including not only teachers / lecturers, but also students, companies, and other experts[55].

There are other associations aiming at networking and exchanging of experience and expertise, like *Regionaltreffen Hilfsangebote Mathematik* (regional meeting for assistance in mathematics), whose member affiliations slightly differ from those listed above. They have either already been mentioned in this section, or they do not meet the filter criteria, i.e. they do not conduct research in mathematics education or do not aim at engineering students.

The three biggest German organisations involved in mathematics education, DMV (Deutsche Mathematiker Vereinigung[56]), GDM (Gesellschaft für Didaktik der Mathematik[57]), and MNU (Mathematisch-Naturwissenschaftlicher Unterricht[58]) founded a commission for the transition from school to university, *Mathematik-Kommission Übergang Schule-Hochschule*. They are involved in the communication with the Ministries of Education in all of the 16 German *Länder* and compose statements concerning the standards of education for mathematics and their implications for schools, universities and the future of mathematics education in Germany. The three societies participating in the commission are themselves involved in assembling and discussing of curricula and educational guidelines. The members of this commission are active in their societies and at their universities and work on various projects in these institutions. The *Kommission Übergang Schule-Hochschule* itself, however, does not conduct any projects of their own.

2.1.6 Engineering courses at RUB

This section covers the specific course conditions at Ruhr-Universität Bochum (RUB). It gives an overview and details on course modules, examination regulations, and the work expected from the students. As such, it provides background to the project undertaking at hand, which must consider local actualities. Whenever these are relevant for the project design, it will be indicated in Chapter 4.

---

[55] http://www.teaching-learning.eu/ueber_uns/ueber_uns.html
[56] dmv.mathematik.de
[57] www.didaktik-der-mathematik.de
[58] www.mnu.de

Courses and course modules

Ruhr-Universität Bochum (RUB)[59] offers different courses in engineering. There is Mechanical Engineering (*Maschinenbau*, MB), Civil Engineering (*Bauingenieurwesen*, BI), Environmental Engineering and Resource Management (*Umwelttechnik und Ressourcenmanagement*, UTRM), Electrical Engineering and Information Technology (*Elektro- und Informationstechnik*, ET/IT), all of which are based on a three-year *Bachelor of Science* course and continue in *Master of Science* courses. There are various foci such as mechanics, energy and process technology, engineering informatics, construction technology and automation engineering, automobile engineering, micro engineering or material engineering, to name but those for the *Master of Science* course in mechanical engineering[60].

Admittance to engineering courses at Ruhr-Universität Bochum has varied in recent years. Due to local changes in school leaving examinations[61], all engineering courses now have locally valid *numeri clausi*. Depending on the average mark in the school leaving exam *(Abitur)*, aspiring students either get a place immediately or have to wait up to three semesters. The requirements for getting a place without waiting are moderate, though, as the average mark[62] necessary ranks from 2.9 (MB) to 3.1 (BI and UTRM) or may even be invalidated as in ET/IT where all applicants were accepted (cf. http://www.ruhr-uni-bochum.de/zsb/nc_werte.ws12.htm, retrieved 08/12/2013). These conditions are easily met by the majority of school leavers. In the years 2010-2012, universities were allowed to dispense with 60% of their places according to their own selection procedure (*Hochschuleigenes Auswahlver-*

---

[59] Ruhr-Universität Bochum was founded in 1965 and is one of Germany's ten largest universities with more than 42,000 students from 130 countries, of which one third study Natural Sciences or Engineering, cf. www.rub.de.

[60] For civil engineering, students can choose between structural engineering, computational engineering, geo technology and tunnel construction, water and environmental management, and traffic engineering. Students aspiring to a degree in UTRM are offered management of processes and products, of energy, of infrastructure and traffic, or of water and soil.

[61] In 2005 North Rhine-Westphalia introduced 12 instead of 13 years of school, meaning increased numbers of school leavers in 2013.

[62] The best possible mark is 0.7, the theoretically possible worst pass mark is 4.0. According to the Ministry for School and Education in North Rhine-Westphalia, average Abitur pass marks deliver an average of 2.50 (in 2012), 2.52 (in 2011) and 2.56 (in 2010) with statistic deviations of 0.67, 0.66 and 0.66 respectively (cf. http://www.standardsicherung.schulministerium.nrw.de/abitur/, retrieved 08/12/2013).

fahren), as long as they gave away 20% of their places to the best students of each year, and 20% to those who had waited long enough. For their 60%, Ruhr-Universität Bochum decided for a mixture of average *Abitur* mark, waiting time, and lottery. In combination with a promotion policy to fill places otherwise left vacant, all this amounted to a rather heterogeneous level of qualifications for students accepted into engineering courses. One vital downside has to be mentioned, though: The complexity of these university entrance procedures not uncommonly resulted in delayed starts, as some students got their admission several weeks after lectures had started. Particularly in the natural sciences and technical subjects, where content builds on content, late starters are severely disadvantaged.

Table 2.9: Overview of Compulsory Courses for the First Semester in MB

Day	Scheduled Lectures	Times
Monday	lecture in Mechanics A, tutorial in Mathematics A, lecture in Mathematics A	10.00 – 18.00
Tuesday	tutorial in Mechanics A, lecture in Basics of Construction Technology, lecture in Materials I	08.00 – 16.00
Wednesday	lecture in Mathematics A, lecture in Physics A	14.00 – 18.00
Thursday	lecture in Mechanics A, lecture in Basic Chemistry, tutorial in Physics	10.00 – 16.00
Friday	practical course in Materials, lecture in Basic Chemistry	10.00 – 18.00

Note. Sessions last 90 – 120 minutes. Times given represent attendance time on campus, free sessions included.

An overview over the compulsory lectures, tutorials and practical courses is given in Table 2.9 (slight variations are possible from one year to another). It amounts to almost 30 hours of lectures and tutorials every week. The lecture and tutorial in mathematics represent a workload of 9 credit points. According to the European Credit Transfer System (ECTS), this translates into a workload of 270 hours per semester (Europäische Kommission, 2009). The lecture and tutorial add up to less than 100 hours, therefore students are expected to do two thirds of their work in mathematics outside of regular university sessions. This is obviously much more than they are used to doing from school. And if this rule were applied to all subjects (not only mathematics), it would mean an immense workload, presumably more than one would expect when starting a university course. In comparison, the OECD average was 1749 hours per year in 2010, and Germans worked 1419 hours per year[63], which translates to 35 to 45 hours per week, depending on the number of working weeks in a year.

All this results in the fact that a considerable number of average or below average students are confronted with demands that not only weigh heavy on their intellectual skills but which also demand the vast majority of their waking hours. It is not surprising that the problems occurring are difficult to overcome.

Examination regulations in mathematics

As mentioned above, there are three different engineering courses whose participants all attend the mathematics lectures MP^2-Math/Plus/Practice focuses on: Mechanical Engineering (*Maschinenbau*, MB), Civil Engineering (*Bauingenieurwesen*, BI) and Environmental Engineering and Resource Management (*Umwelttechnik und Ressourcenmanagement*, UTRM). For all three, the lecture *Mathematics 1* (*Higher Mathematics A* for BI and UTRM) is a required course in the first semester and has to be followed up by *Mathematics 2* and *Mathematics 3 and Numerical Mathematics* (*Higher Mathematics B* and *Higher Mathematics C and Basics in Numerical Mathematics*, respectively) in the subsequent semesters.

All students are registered automatically for the written exams in mathematics (Ruhr-Universität Bochum, 2009a, 2009b, 2009c, §5 (14), §5 (11), §5 (11) respectively) and cannot unregister (§5 (18), §5, (17), §5 (17)). If they fail an attempt, students are automatically registered for the next possible regular examination date (§5 (17), §5, (15), §5 (15)). What is more, all students only have the limited number of four attempts in order to pass the written exam (§5 (20), §12 (1), §12 (1)), plus an oral examination which means the restriction of a mere pass grade. Each failed attempt counts, non-attendance counts as failure (§5 (22), §5, (16), §5 (16)). Students are automatically registered for the mathematics lectures of the second and third semester even if they have not passed the first semester yet. Similar regulations apply for other obligatory first

---

[63] http://dx.doi.org/10.1787/888932505564

year courses, such as Physics, Mechanics and Chemistry (depending on the exact engineering course attended) which themselves are regarded as serious obstacles.

The written examination in *Mathematics 1* or *Mathematics A* respectively itself consists of 10 to 12 exercises, graded in a points system. It is possible to include multiple choice exercises, too, so some lecturers do so as it relieves the strain of grading several hundred papers. In order to avoid passing by coincidence, in some years minus points for wrong answers in the multiple choice questions were introduced. To get a pass grade, students have to accumulate a minimum of 50% of the points possible at the written exam. The bonus points acquired before with the help of homework or mini exams count for a little extra. Nevertheless, the pass rates have a longstanding tradition of around 50%[64]. The lecturers in mathematics and the department of engineering have decided that no technical tools whatsoever are allowed, meaning that in spite of working with graphing calculators at school, students have to compute everything by hand at university.

All these regulations mean that students must come to grips with their workload in their first year at university, otherwise they face the danger of having to cope with twice as much in their second year. As the official workload for a mathematics course is 270 hours per semester, of which two thirds are to be spent in private study (see section 2.1.6), this borders on the impossible.

The aim of these regulations is, naturally, to put pressure on students to apply themselves, to work hard, and to not waste time. The rules also discourage pro forma students who only enrol but do not attend courses, as they will quickly collect failed attempts and therefore must leave the course.

Lectures, homework, and tutorials

As depicted in Table 2.9, in *Mathematics 1* respectively *Mathematics A* the students are expected to attend four hours of lectures every week and two hours of tutorials. The lectures are held in lecture halls which seat more than

---

[64] According to Heublein, Schmelzer, and Sommer (2008); Heublein et al. (2012); Heublein, Richter, Schmelzer, and Sommer (2014), the dropout rate in engineering courses varies between 30% and 50% (meaning graduation rates between 50% and 70%), depending on the type of university, with a decreasing tendency. It is therefore legitimate to estimate the pass rates of single examinations lower.

Table 2.10: Contents of Mathematics Lecture in the First Semester

1 Sets, Numbers and Functions [✓]	9 Linear Algebra
2 Vectors ✓	10 Limits [✓]
3 Straight Lines and Planes ✓	11 Elementary Functions ✓
4 Systems of Linear Equations ✓	12 Differentiation ✓
5 Matrices [✓]	13 Applications of Differentiation ✓
6 Determinants	14 Integration ✓
7 Eigenvalues and Eigenvectors	15 Applications of Integration [✓]
8 Quadrices	16 Differential Equations

Note. ✓: usually covered in depth at school, [✓]: covered superficially

800 students with the help of microphones and screen presentations (individual lecturers may prefer chalk and blackboard). They follow the traditions of mathematics lectures in so far as the lecturers usually present the material from an axiomatic point of view (see section 2.1.2). In mathematics lectures for engineering students, however, proofs are often omitted and instead the stress is put on examples and applications relevant for engineering. The lectures are complemented by a printed (or printable) script and subsequent uploads of the transcriptions from the presentations during the lectures. And of course there are several books around which especially cater for engineering students and their demands on mathematics, e.g. Papula (2011), Papula (2010) or Meyberg and Vachenauer (2003).

The depth of the mathematics taught to engineering students in their first year, however, is reasonable. Several branches are also taught at school and students should, theoretically, be familiar with the concept of derivation and integration. The themes are shown in detail in Table 2.10[65], together with indications if they are commonly dealt with at school. In an ideal world, a considerable part of the content of the first semester would not really be new for the students. In reality, however, the school curriculum allows variations, different foci and omissions, owing to the homogeneity accepted in modern classrooms.

There is homework, usually four exercises per week which can be handed in in order to be graded which in turn can lead to a few bonus points counting

---

[65] http://www.ruhr-uni-bochum.de/imperia/md/content/mathematik/service-zentrum/skripte/ mbbi1_2012.pdf, pp. 2-3

for the final written examination. Additional analogous exercises are discussed in the tutorials, in comparably small groups of no more than 40 students, in compliance with constructivist learning theories, see section 2.1.2. The tutorials are held by teaching assistants, research assistants or senior students. The steps in the solutions are mostly explained in detail and more slowly in the tutorials than would be possible during the lecture. The majority of students prefer to work on their mathematics homework in groups, at least at some stage during the solution process. This is encouraged by the fact that it is traditionally allowed to hand in homework in groups of up to three students. Although this may also foster mindless copying of exercises, it also opens the perspective for fruitful cooperative learning. Both the engineering and the mathematics department have rooms available in their libraries and cafeterias where students can sit together and discuss their tasks and homework. Hence many of them spend a good part of their time on campus, even outside lecture times.

It has become a custom, too, to have mini exams two or three times during the semester. These usually last 15 to 20 minutes and are held at the beginning of a lecture. They consist of four short exercises, sometimes multiple choice, and cover the lecture contents of three to four weeks. Students are granted bonus points that count for the final examination. In order to prevent cheating in the crowded lecture halls, there are up to three different sets of tasks. The mini exams are graded by the teaching assistants and discussed in subsequent session of the tutorials. The mini exams are regarded as serious tests as they are written under similar conditions (independent performance, no calculators or other technical tools) and are therefore the starting point of MP^2-Math/Plus, see chapter 4.3.1.

Some lecturers additionally offer collective tutorials where solutions for the weekly homework tasks are presented. This may differ from one year to the other according to attendance and personal resources. The number of uploads in the accompanying e-learning course also varies substantially, depending on the preference of the lecturers and the time on their hands. A perfectly-elaborated example is the year 2012/2013 where every single lecture was accompanied by a pre-learning video of five to seven minutes which introduced the topic of the oncoming lecture and ended with a multiple choice question which again awarded some bonus points for the final examination.

The university's SZMA helpdesk provides extra support for students attending a course in service mathematics seeking specific support: The department of mathematics at RUB has been aware of the problems students from other departments face when they attend lecture in mathematics, and therefore founded a *Service Centre for Mathematics and Applications Servicezentrum für Mathematik und Anwendungen*, SZMA in 2007[66], which offers help and support. Among other things, the SZMA offers training courses for tutors and a helpdesk. This helpdesk is open three hours every weekday afternoon[67] and staffed by teaching assistants, tutors and senior students employed for marking homework. Students can visit the helpdesk without an appointment (preferably when someone associated with their mathematics lecture is on duty) and get their questions answered.

The conditions described boil down to the fact that the demands in mathematics on engineering students are considerable – particularly when it comes to choosing the modalities which work for them and to studying independently.

2.2 Learning Strategies

This section is part of the broad Chapter 2 which contains the theoretical background relevant for our study. It presents the information on learning strategies, as the last step before the research approach and objectives are addressed in Chapter 3.

When aiming at an improved learning output, it is imperative to look closely at the processes of learning, i.e. of building competence. In agreement with e.g. Wild (2005) and Rach and Heinze (2011), a promising perspective is recognised in exploring general and meta-level skills in terms of learning strategies, whose investigation allows for revealing both the cognitive dispositions as well as affective barriers and pathways – and finally the interrelations between them. As shown in section 2.1.5, numerous research and instruction projects that attend to these issues can be found in Germany alone (for an overview of international projects, see Dunn et al., 2012). The causality between adequate learning strategies and successful learning seems well established (for example cf. Erdem Keklik & Keklik, 2013), although the perspective of how to sustainably

---

[66] http://www.ruhr-uni-bochum.de/ffm/szma/
[67] The opening times were expanded to four hours every weekday in 2013.

encourage learning strategies, to identify justified combinations of interventions, and to understand the influence of motivation needs to be further explored.

Apart from a talent for abstract thought and formalities, mastering a university course containing mathematics needs a combination of broader capacities, like general skills and attitudes such as self-organisation, perseverance and frustration tolerance, as well as subject-specific abilities (cf. Pintrich, Smith, Garcia, & McKeachie, 1993; Weinstein & Palmer, 2002; Wild, 1994). It is students' meta-level learning behaviour[68] that is crucial, taking account of the words of de Guzmán et al. (1998), who state:

> Students' success is linked to a great extent to their capacity of developing "meta-level" skills allowing them, for instance, to self-diagnose their difficulties and to overcome them, to ask proper questions to their tutors, to optimise their personal resources, to organise their knowledge, to learn to use it in a better way in various modes and not only at a technical level (p. 760).

Thus, when looking for a suitable catalogue of learning strategies, it is essential to search for a comprehensive inventory that includes these skills. Accordingly, learning strategies are understood as all kinds of planned and conscious learning behaviour and the attitudes behind it, involving observable actions (e.g. solving tasks, asking questions, taking notes) as well as thought processes (e.g. planning, reflecting) on the basis of both cognitive and affective-motivational dispositions[69]. This perspective is supported by Blömeke, Gustafsson, and Shavelson (2015) who contribute to modelling competence "as a process, a continuum with many steps in between" (p. 7). In particular they emphasise the following approach:

> Thus, we suggest that *trait* approaches recognize the necessity to measure behaviorally, and that *behavioral* approaches to competence recognize the role of cognitive, affective and conative resources. At this time, we encourage research on competence in higher education emanating from either perspective and paying attention particularly to the steps in between (p. 7).

---

[68] However, de Guzmán et al. (1998) do not mean metacognition in the sense described in section 2.1.2 when characterising meta-level learning behaviour, but rather learning strategies in terms of Wild (2005), see Figure 2.9.

[69] This extends to a lack of planning and conscious actions as well, as it also presents a characteristic of an individual's learning strategy.

As depicted in section 2.1.2, research on the significance of learning strategies in mathematics education has its roots in contributions highlighting the
role of affect, motivation and beliefs (McLeod, 1992; Hannula, Evans, et al.,
2004; Leder et al., 2002; Leder & Grootenboer, 2005; Roesken & Casper,
2011; Zan et al., 2006), as all cognitive processes involve affective stances
that moderate the tension between modes of intuitive and analytical thinking
(e.g. Fischbein, 1987; Stavy & Tirosh, 2000). It can be said, therefore, that
this has led to a fortified interest in certain kinds of learning strategies. In
the context of mathematics, overcoming motivational and affective barriers
with the help of meta-skills, e.g. self-regulation, has become an important
issue. What is more, mathematics demands the use of effective planning
as well as organised and consistent work (cf. Rach & Heinze, 2011). More
than many other subjects, mathematics is cognitively challenging and needs
motivational perseverance, thus representing an ideal research area for the
influence of interventions addressing learning strategies, both on a general
and a meta-level. The goals when assessing students' learning behaviour are
various, however: taking an inventory, describing the development, comparing
or improving learning behaviour (cf. Lovelace & Brickman, 2013), and so are
the research interests, in mathematics popularly performance prediction or the
identification of at-risk students. Apart from more time-consuming methods (like
observing groups of students or individual students when learning mathematics,
using thinking-aloud and videotaping, or keeping track of the learning process
by convincing learners to write detailed learning logs), this has led to a great
variety of questionnaires in many languages.

2.2.1 Questionnaires for assessing learning strategies

Questionnaires with different focal points (according to the background of the
authors and their research interests) originate from this variety (cf. Pintrich et
al., 1993; Weinstein & Palmer, 2002). Schellings (2011) gives a comprehensive
overview from an international and a Dutch perspective. Though her work is
based on the text-heavy learning of history, the general categories of learning
behaviour can be applied to other subjects as well. Differentiating between
motivational and cognitive aspects when dealing with learning strategies is
a widely accepted concept (cf. Nenniger, 1999) and is in keeping with the
understanding of affective aspects as a key issue.

In the following, approaches to capture learning strategies which have influenced subsequent research fundamentally will be outlined. The selection includes only those which reflect the importance of affective and motivational issues. Pintrich et al. (1993) developed a questionnaire "to measure college undergraduates' motivation and self-regulated learning" (Artino, 2005, p. 3), the Motivated Strategies for Learning Questionnaire (MSLQ). The MSLQ measures motivation and self-regulated learning in general and for a particular course by means of six motivation and nine learning strategies subscales. Initially, Pintrich and de Groot (1990) started by postulating a five latent factor structure comprising expectancy, value, affect, learning strategies, and self-regulation. The items that were developed for operationalising these constructs later formed the basis for the 15 subscales (six for motivation, nine for learning strategies) mentioned above. MSLQ has been applied in many research studies (Duncan & McKeachie, 2005), partly aiming at developing a new conceptualisation with respect to the significance of the single sub-scales (Dunn et al., 2012; Hilpert, Stempien, van der Hoeven Kraft, & Husman, 2013). MSLQ's reliability has turned out to be "robust" and its predictive validity to actual course performance is considered "reasonable" (Pintrich et al., 1993, p. 801).

The Approaches to Studying Inventory (ASI) by N. Entwistle and Ramsden (1983) and its refinements (ASSIST by Tait, Entwistle, & McCune, 1998, ALSI by N. Entwistle & McCune, 2004) feature the main distinction of categorising learning behaviour as being of either strategic (deep) or of apathetic (surface) approach. The dichotomy forms the inventory's two main factors which in turn contain up to 16 subscales, depending on the version of the questionnaire. Although the authors do not group their items into motivational and (meta-)cognitive scales, the object of research is nearly identical to that of MSLQ users. A specific feature of ASI and its variations is the idea to measure not only the desired learning behaviour (strategic approach), but also what is hypothesised as less success-oriented (apathetic approach). This produces a multifarious picture of learning behaviour.

Another well-known instrument to capture students learning strategies is the Learning and Study Strategies Inventory (LASSI) by Weinstein and Palmer (2002). LASSI covers thoughts, behaviour, attitudes and beliefs in relation to successful learning that can also be fostered by interventions. Its ten scales are classed into affective strategies, goal strategies, and comprehension monitoring strategies, thus covering cognitive, metacognitive (particularly self-regulative),

affective and motivational aspects. LASSI is not only used for research purposes, but is also recommended to students to use for themselves in order to
get feedback on their strengths and weaknesses. LASSI's reliability coefficients
(Cronbach's α) for its different scales are reported to score between .86 and .68,
the lowest often being considered insufficient (Weinstein, Schulte, & Palmer,
1987). Its validity to academic performance depends on the specific scale, e.g.
Cano (2006) found, using multiple regression, that two scales (namely *Affective
Strategies* and *Goal Strategies*) contributed to academic performance, whereas
one (*Comprehension Monitoring Strategies*) did not.

All questionnaires described so far resort to self-assessment of student
behaviour. It must be conceded that this entails the weakness that "the learner's
perceptions of his or her strategies are measured" (Schellings, 2011, p. 94),
which need not coincide precisely with the strategies themselves. In this context
it is interesting to compare self-reported learning behaviour (especially concerning metacognition) to the results gained with other methods, e.g. *Thinking Aloud*
(David, 2013). Considering affective aspects, however, the learner's subjective
perspective is what counts. Other problems, like assessing the sufficiency or
efficiency of study time or effort, might be harder to overcome.

2.2.2 Origin and structure of the LIST questionnaire

For our research at Ruhr-Universität Bochum in Germany, the decision fell
for the German LIST questionnaire (Learning Strategies at University, Wild
and Schiefele, (1994)), which is based on the same classification as MSLQ
and takes up aspects from LASSI as well. LIST was invented for measuring
learning strategies of medium generality, between learning styles and learning
tactics (Wild, 2000). The instrument distinguishes between cognitive, metacognitive and resource-related learning strategies and comprises dimensions of
learning strategies grouped accordingly. This mirrors the acceptance of this
taxonomy in the German-speaking community (Wild, 2000). Just like its English
predecessors, this approach originates from educational research and thus is
not subject-specific. However, in university mathematics education the instruments are frequently used to assess students' learning behaviour on a general
level while combining the results with subject-related measures (Lovelace &
Brickman, 2013).

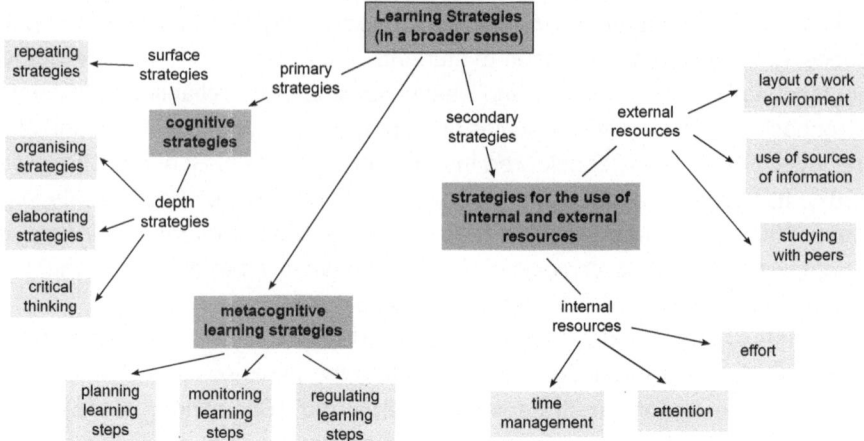

Figure 2.9: Overview of Learning Strategies (Wild, 2005, p. 194), translation by
author

LIST covers various learning strategies, which are divided into cognitive,
metacognitive, and resource-oriented scales, see Figure 2.9. According to Wild
(2005), cognitive (primary) learning strategies encompass surface learning
(repeating, rote-learning) as well as deep learning strategies (e.g. re-organising
of and elaborating on subject matter[70]). Metacognitive learning strategies cover
planning, monitoring, and regulating the next steps in the learning process.
Resource-oriented learning strategies are subdivided into those using external
resources (e.g. sources of information, working with peers) and those employing
internal resources (effort, attention, and time management). The version of
LIST employed in our research can be found in Appendix A, page 239, along
with an English translation in Appendix B, page 254, see also Table I.1 on page
280.

The LIST questionnaire for measuring learning strategies in academic
studies was first compiled in the 1990s (Wild & Schiefele, 1994) and has since
been modified and tested several times. LIST has been applied in the context
of many subjects, mathematics among them (cf. Liebendörfer et al., 2014, for
an overview), with overall satisfying results with regard to reliability (Wild and
Schiefele (1994) found Cronbach's α between .64 for *Metacognition* and .90

---

[70] Particularly elaborating strategies are regarded as central for the learning of mathematics, see
Göller et al. (2013).

for *Attention*) and validity (Wild, 2000, 2005). In the following, how LIST takes up scales and items from MSLQ and LASSI is explored in detail, in order to understand its origins and to illustrate its structure.

Apart from *Motivation*, the scales from LIST are derived directly from MSLQ, although the number of items varies. Some items in LIST are translations of MSLQ items. In comparison to LIST, MSLQ seems very differentiated in terms of motivation, it incorporates six *Motivation* scales (*Intrinsic Goal Orientation*, *Extrinsic Goal Orientation*, *Task Value*, *Control of Learning Beliefs*, *Self-Efficacy for Learning and Performance*, *Test Anxiety*) comprising 31 items. LIST does not have items with the label *Motivation* as such, but LIST's six items on *Attention* (which are all reverse coded) and eight items on *Effort* more or less cover this aspect, for example "I work late at night or at the weekends if necessary". And other LIST scales, in particular the resource-oriented ones, are meant to measure the degree of motivation a student possesses when preparing for an important exam, with items like "I fix the hours I spend daily on learning in a schedule". The main difference between the two questionnaires is that MSLQ puts more emphasis on including different aspects of motivation as *Goal Orientation*, or *Control of Learning Beliefs*. For LIST, on the other hand, the aim was to clearly keep apart cognitive and motivational aspects.

LASSI (Weinstein & Palmer, 2002) also separates cognitive aspects, but has much less communalities with LIST. LASSI scales partly cover the same contents though holding different names, e.g. *Concentration* and *Attitude* (LASSI) compared to *Attention* (LIST). The numbers of items in a scale are different, too: There are 3 to 8 in LIST (if *Metacognitive Strategies* are divided into three scales, 4 to 8 if not), 3 to 12 (3 to 8) in MSLQ, and a constant 8 items in all LASSI scales. This results in considerable differences in analogous scales between LIST and MSLQ: LIST has 31 in *Cognitive Strategies* whereas MSLQ has 19 in the respective scales. According to the inventors of LIST, scales were expanded in order to reach better reliability (Wild, 2000). All three questionnaires use Likert scales, ranging from 5 points (LIST) over 6 (LASSI) to 7 (MSLQ). An overview on how LIST is based on MSLQ and LASSI is provided in Table 2.11.

Table 2.11: Synoptic Table for LIST, MSLQ, and LASSI

	LIST		MSLQ		LASSI
Cognitive Strategies	Critical Checks (8 items)	**Learning Strategies**	Critical Thinking (5 items)	**Affective^A Strategies, Goal^G Strategies, Comprehension^C Monitoring Strategies, see also below**	
	Elaborating (8 items)		Elaboration (6 items)		Information Processing^C (8 items)
	Organizing (8 items)		Organization (4 items)		Selecting Main Ideas^G (8 items)
	Repeating (7 items)		Rehearsal (4 items)		
Resource-related Strategies	Attention (6 items)		Motivation* (see below)		Attitude^A (8 items)
	Effort (8 items)		Effort Regulation (4 items)		Concentration^A (8 items)
	Learning Environment (6 items)		Time / Study Environmental Management (8 items)		
	Time Management (4 items)				Time Management^A (8 items)
	Peer Learning (7 items)		Peer Learning (3 items)		
	Using Works of Reference (4 items)		Help Seeking (4 items)		Study Aids^C (8 items)
Metacogn. Strat.	Planning (4 items)		Metacognitive Strategies (12 items)		Self-Testing^C (8 items)
	Monitoring (4 items)				
	Regulating (3 items)				

A = affective strategies, G = goal strategies,
C = comprehension monitoring strategies

Synoptic Table for LIST, MSLQ, and LASSI, continued.

LIST		MSLQ		LASSI
Resource-rel. Strat.	Attention (see above)	Motivation	Control of Beliefs* (4 items)	Affective Strategies[A], Goal Strategies[G]	
			Self-Efficacy for Learning and Performance* (8 items)		
			Intrinsic Goal Orientation (4 items)		Motivation[A] (8 items)
			Extrinsic Goal Orientation (4 items)		
			Task Value (6 items)		
			Test Anxiety (5 items)		Anxiety[G] (8 items)
					Test Strategies[G] (8 items)
	77 items		81 items		80 items

A = affective strategies, G = goal strategies

3 Research Approach and Objectives

In Chapter 1 the problem of the high dropout rates of engineering students was outlined, followed by an extensive description of the theoretical concepts relevant for understanding it in its complexity (Chapter 2). In this chapter, these facts and theories are shaped into a research approach, that the project design (Chapter 4) and the empirical evidence (Chapter 5) can refer to, see orientation figure below.

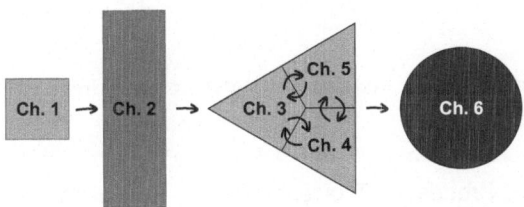

The present chapter considers the research concerns and objectives and deduces conditions for a suitable approach (section 3.1). Then the actual project specifications are revisited in front of this background (section 3.2), and the research questions are phrased (section 3.3).

3.1 The *Design Research* Approach

Our research concerns mathematics for first-year engineering students. This group of students may not be interested in mathematics as a formal-axiomatic world (according to Tall, 2004), but as a means to solve practical engineering problems. The purpose of our research project is to support them to cope with the quantity and quality of mathematical contents they have to master in a given time. From appraisal of the literature on learning mathematics, it seems well-advised to design a learning environment that enables students to involve themselves with the subject matter in context (Dreyfus, 2012) and to discuss their approaches with peers and teachers (von Glasersfeld, 1991), as well as to address selected affective aspects, namely beliefs, attitudes, and approaches to learning, see section 2.1.2 (pages 30 to 35). The choice of methods should include sufficient resources and alternative scenarios of the learning process (hence learning strategies), and likewise treat the learner as an autonomous being in control of possible outcomes, see section 2.1.2 (particularly page 42). This suggests first design ideas which will be elaborated upon in Chapter 4.

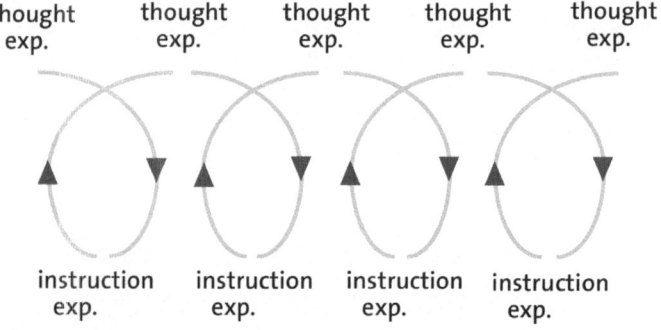

Figure 3.1: Cyclic Processes Typical of *Design Research*,
(Gravemeijer & Cobb, 2013, p. 81)

There are two kinds of objectives: Practically, the objective is to design, conduct, evaluate, and improve interventions. Theoretically, the objective is to understand which interventions work and why that is the case. The kind of interventions in mind should preferably also be practical (because they are meant to last), repeatable (because they will be used again every year), and transferable (because other universities face similar problems). These cues prompt a research approach that encompasses these three aspects.

Design Research[1] (Gravemeijer & Cobb, 2013) regards teaching and its reflection as cyclic processes of thought and instruction experiments, see Figure 3.1. What will be called *Design Research* here has had (and still has) other denotations, e.g. design experiments, design-based research, design studies, developmental research or engineering research, to name but a few.

---

[1] For a concise overview on the origins and propagation of *Design Research*, see Prediger, Gravemeijer, and Confrey (2015).

The researcher to be credited as setting *Design Research* in motion is Brown (1992) who more than 20 years ago demanded that researchers

> must operate always under the constraint that an effective intervention should be able to migrate from our experimental classroom to average classrooms operated by and for average students and teachers, supported by realistic technological and personal support (p. 143).

Traditional experimental designs do not suit this focus well. Numerous researchers criticise educational research in more than one respect. The Design-Based Collective (2003) declare that "educational research is often divorced from the problems and issues of everyday practice" (p. 5). Brown (1992) finds that "Aspects of it [classroom life] that are often treated independently [...] actually form part of a systemic whole" (p. 141f.), whereas traditional educational research often concentrates on a single isolated phenomenon. Barab and Squire (2004) claim that this kind of research also regularly takes place under laboratory conditions, outside normal classrooms or lecture halls, where children or students are mentored by unfamiliar people in an unfamiliar setting. Collins (1992) finds fault with the fact that there is frequently only little cooperation between instructors and researchers, he puts "teachers as co-investigators" (p. 16) at the top of his list of desiderata. These critiques have led to an increase in *Design Research* publications, as particularly in the last ten to fifteen years, many researchers and educational reformers have felt like Barab and Squire (2004), Brown (1992), Collins (1992), and the Design-Based Collective (2003) and have thought of a generalisable way to conduct their research. As Anderson and Shattuck (2012) show in detail, the number of publications on *Design-Based Research* has steadily grown between the years 2000 and 2010, although (in agreement with McKenney and Reeves, (2013)) their method of gaining these numbers must be seen critically[2]. All approaches in this line comply with certain qualities, as van den Akker, Gravemeijer, McKenney, and Nieveen (2006) summarise:

 – Interventionist: the research aims at designing an intervention in the real world;

---

[2] For example, there are no numbers as to how comparable areas of research have grown, how much digitalisation has influenced online searchability of research papers, and how many publications were ruled out because of the fact that they had appeared in books or conference proceedings.

- Iterative: the research incorporates a cyclic approach of design, evaluation, and revision;
- Process oriented: a black box model of input-output measurement is avoided, the focus is on understanding and improving interventions;
- Utility oriented: the merit of a design is measured, in part, by its practicality for users in real contexts; and
- Theory oriented: the design is (at least partly) based upon theoretical propositions, and field testing of the design contributes to theory building. (p. 5)

Plomp (2013) suggests, in reference to other authors from the field (van den Akker, Branch, Gustafson, Nieveen, & Plomp, 1999; Kelly, 2006; Nieveen, 1999), adding another characteristic:

- Involvement of practitioners: the research involves active partici- pation with practitioners in the various stages and activities of the research – this will increase the chance that the intervention will indeed become relevant and practical for the educational context which increases the probability for a successful implementation. (p. 20)

The research approaches that fulfil these characteristics all stem from the desire to conduct research that has a measurable impact. It is no surprise that many researchers (van den Akker, 1999; Reeves, 2006; Barab & Squire, 2004; Design-Based Collective, 2003) have developed similar approaches under varying names, for the three main motives that van den Akker et al. (2006) give as "the desire to *increase the relevance* of research for educational policy and practice", the aim "at *developing empirically grounded theories*", and "the aspiration of *increasing the robustness of design practice*" (p. 3f., emphasis in the original). These motives also hint at the progenitors of the approach: academic researchers whose expertise is requested whenever new educational policies are about to be implemented. The underlying vision is to realise sustainable changes in the educational system that are based on sound research. Of course, this objective is not exclusive to *Design Research*, but "In design research, a theorist or researcher's rigorous analysis of a learning problem leads to quite specific ideas for interventions" (Walker, 2006, p. 9).

The "two main outputs from educational design research: maturing inter- ventions and theoretical understanding" (McKenney & Reeves, 2012, p. 80) are not always produced in balance, though. This is why some researchers have made allowance for *Design Research* projects that have a stronger emphasis on developing interventions, and those that more explicitly aim at advancing theory. Plomp and Nieveen (2013a, p. 12, p. 16ff.) and Nieveen, McKenney, and van den Akker (2006, p. 152f.) term them "developmental studies" and "validation studies"; McKenney and Reeves (2012, p. 23) supply the respective descriptions "research *through* interventions" and "research *on* interventions" (emphasis in the original). According to Plomp (2013), the two types of studies[3] can be defined as follows.

> In the case of **development studies**, the purpose of educational design research is to develop research-based solutions for complex problems in educational practice. This type of design research is defined as *the systematic analysis, design and evaluation of educational interventions with the dual aim of generating research-based solutions for complex problems in educational practice, and advancing our knowledge about the characteristics of these interventions and the processes of designing and developing them.*
>
> On the other hand, in **validation studies** the purpose of design research is the development or validation of a theory, and this type design research is defined as *the study of educational interventions (such as learning processes, learning environments and the like) with the purpose to develop or validate theories about such processes and how these can be designed.* (p. 16, emphases in the original)

Other researchers (e.g. Prediger et al., 2015) maintain that "generating theories is crucial for design research" (p. 877), which has become a popular buzzword. This brings up a more basic problem, to define what a theory is exactly. If you content yourself with the notion that a theory (in educational science) "advance[s] our knowledge about the characteristics of [...] interventions and processes to design and develop them" (Plomp, 2013, p. 15), developmental studies can be classified as *Design Research*. In this context, Burkhardt (2006, p. 131) points out the fact that "Education is a long way behind [...] in the range and reliability of its theories", meaning the "status and role of theory" (Burkhardt, 2006, p. 130) have to be taken into consideration when demanding

---

[3] In educational *Design Research*, there is a third type, implementation studies, which aim at up-scaling a curricular innovation and exploring conditions for its successful realisation.

theoretical output. This is in line with Phillips (2006) who invokes tolerance for the "many different frameworks and methodological approaches" (p. 94) when judging a *Design Research* project proposal. Rejecting a thought-out proposal or denying the validity of thoroughly investigated results of such a project entails more disadvantages than benefits, as "Design research offers the opportunity to create successful innovations and learn lessons that cannot be achieved through design and empirical research independently" (Edelson, 2006, p. 105).

All *Design Research* studies are organised in three phases: analysis of the problem, design of the intervention, and evaluation of the design (McKenney, Nieveen, & van den Akker, 2006), which is repeated in iterative cycles. Plomp (2013) uses different terms: preliminary research, development or prototyping phase, and assessment phase (p. 30), see descriptions in Table 3.1. Complementing the notion of Gravemeijer and Cobb (2006) as presented in Figure 3.1, she both concretises the design process and puts it in an elevated chronological sequence. The thought and instruction experiments happen during these phases as well as in between and spanning them.

Table 3.1: Phases in *Design Research* Projects

Phase	Short Description of Activities
Preliminary research	Review of the literature and of (past and / or present) projects addressing questions similar to the ones in this study. This results in (guidelines for) a framework and first blueprint for the intervention.
Development or Prototyping phase	Development of a sequence of prototypes that will be tried out and revised on the basis of formative evaluations.
Assessment phase	Evaluate whether target users can work with intervention (actual practicality) and are willing to apply it in their teaching (relevance and sustainability). Also whether the intervention is effective.

Note. Adapted from Plomp (2013, p. 30).

Researchers in *Design Research* are well aware of the challenges of this approach. The most prominent is: "The researcher is designer and often also evaluator and implementer" (Plomp, 2013, p. 42). This can hardly be avoided, particularly in a close project team where everyone puts in their expertise. What helps alleviating conflict is, among others, open discussion, critical friends, triangulation of various sorts (concerning data and investigators), systematic documentation, and empiric evidence (see Plomp, 2013, p. 42). Another recommendation (Plomp, 2013) seems both useful and practical: "shift from a dominance of 'creative designer' perspective in the early stage, towards the 'critical researcher' perspective in later stages" (p. 42). Other challenges in *Design Research* are "Real-world settings bring real-world problems" and "Adaptability of the research design" (Plomp, 2013, p. 42f.). The first is self-evident, and may be overcome with the help of the practitioners in the project team. The last refers to the flexibility of project interventions, the commitment of those involved to tackle challenges when they arise, and to doubt and discard what they believed in at first – in sum, it boils down to a competent project team and the working atmosphere created and maintained by the project management.

3.2 Project Conditions

We have seen that a *Design Research* approach fits our project purposes, and promises to lend background and structure on which to model further project design and evaluation. But does MP^2-Math/Plus/Practice meet the requirements of *Design Research*?

The idea for MP^2-Math/Plus/Practice was conceived from the experience that the traditional way to teach mathematics to engineering students did not seem to accommodate a considerable share of students. The aim from the very beginning was to make an impact on their learning of mathematics, with the ambitious goal of preventing unnecessary drop-out. The expressed aim of MP^2-Math/Plus/Practice was to make a real difference for those engineering first-year students who need support in order to master their mathematics lecture – and hopefully their engineering course as a whole. This is the intention that triggered *Design Research*: to aim at an **intervention in the real world** (see van den Akker et al., 2006, p. 5, as cited here on page 79).

The second demand for **iteration** fits the organisation of MP^2-Math/Plus/ Practice perfectly as it is active only in winter semesters (Math/Plus) respectively summer semesters (Math/Practice), developed by different teams. This allows time for analysis, evaluation, and eventual reconception in between project cycles. The iterative nature of MP^2-Math/Plus/Practice was clear from the start, when the project was granted funding for three years. The prospect of continuation in case of success was in view from the beginning, as it seemed highly probable that the problem (a considerable number of first-year engineering students with difficulties in mathematics) would persist. The descriptions of the three phases matches the strategy in MP^2-Math/Plus/Practice, where the preliminary phase took place before the project conception, and the development phase and assessment phase are repeated at least once in each project cycle, followed by an overall assessment and evaluation in the form of this thesis.

The project management recruited experts from different fields, including mathematicians with experience in teaching engineering mathematics, a mathematics educator / researcher in mathematics education, a computer scientist, project managers, an educationalist, and an experienced school teacher. The relatively small size of the team and the fact that they were located in the same department at the same university facilitated communication, even enhanced by the fact that about half of them had already worked together. The broad expertise meant that aspects could be discussed under various angles, with the intention of understanding what went wrong, what went right – and why. This meets the *Design Research* requirement of **process-orientation**. What is more, with a background in mathematics, no-one in the project team was prepared to simply accept the state of affairs, everyone wanted to understand the reasons why (relatively simple) mathematics poses such a huge problem for engineering first-years.

The **orientation towards utility** was always on the agenda, as the intervention was meant to take place where university mathematics courses are usually situated, in the department of mathematics. The rooms, the time slots, the size of the groups, the ways of communication etc. are all standard for tertiary education tutorials. From the beginning there was the idea to upscale the intervention in case it went well, which lent even more importance to the fact that the interventions had to be practical in different real-world contexts, at other universities under slightly different conditions. With sustainability in mind, the interventions were designed with an eye on how to repeat them with less expenditure, by ourselves in our own university of Ruhr-Universität Bochum (RUB), by others (research assistants, teaching assistants, and student assistants) at RUB, as well as at other universities.

Involving practitioners was an idea from the start, too. The size of the project team would not have allowed for a great distance between practitioners and researchers, and it was never considered a criterion. As mentioned in section 3.1, closeness between practitioners and researchers can be regarded as a problem, see McKenney et al. (2006, p. 83) who elaborate on the common phenomenon that the designer is both "implementer and evaluator". The remedies suggested (exchange, open discussions, constructive criticism, triangulation, systematic documentation, use of empiric evidence, shift from designer to researcher over time) were put in regular use.

It is **orientation towards theory** that is not satisfied by MP^2-Math/Plus, however, if you define it as contributing to new theories (or refining established ones) on the learning of mathematics. As MP^2-Math/Plus does not have a focus on the development of theories, it falls under the label of development studies, and it fits this label well. Its pronounced aim is to design (a combination of) interventions for the complex educational problem of teaching mathematics at tertiary level, and to evaluate them, which brings with it an advancement "of knowledge about the characteristics of these interventions and the processes of designing and developing them" (Plomp, 2013, p. 16).

As MP^2-Math/Plus/Practice meets the requirements of a *Design Research* project to this extent, general quality criteria should be considered. According to Plomp and Nieveen (2013b) who draw on Nieveen (1999),

> a complete and final version of an intervention should be
>
> - *Relevant*: There is a need for the intervention and its design is based on state of the art (scientific) knowledge – also called *content validity*;
>
> - *Consistent*: The intervention is 'logically' designed – also called *construct validity*;
>
> - *Practical*: The intervention is usable in the setting for which it has been designed;
>
> - *Effective*: Using the product results in desired outcomes
>
> (p. VIII, emphases in the original)

This is what the outcome of the work on MP^2-Math/Plus/Practice has to measured upon. The first point, relevance, was covered in Chapter 1. The second and third, consistency and practicality, will be elaborated on in detail in the next Chapter 4. The last criterion, effectiveness, needs more preparation (Chapter 5) and will finally be addressed in Chapter 6, notwithstanding the general guiding principles for scientific research in education (Shavelson & Towne, 2002) that everyone must respect, namely

> – Pose significant questions that can be investigated empirically.
> – Link research to relevant theory.
> – Use methods that permit direct investigation of the question.
> – Provide a coherent and explicit chain of reasoning.
> – Replicate and generalize across studies.
> – Disclose research to encourage professional scrutiny and critique.
>   (p. 52)

3.3 Research Questions

The situation demands much from MP^2-Math/Plus/Practice interventions which are expected to produce an impact despite very limited time, to help students overcome organisational obstacles when learning, to take different educational backgrounds into consideration and still have the same objective in mind for everyone – and to exhibit an affective component. Planning a research project under these demands and conditions can be traced back to van den Akker (2013) who generalises his findings of curricular *Design Research* as follows.

> These [design] principles may be captured in (a growing set of) heuristic statements of a format such as:
>
> – *If you want to design intervention X [for purpose/function Y in context Z]*
> – *then you are best advised to give that intervention the characteristics C_1, C_2, ..., C_m [substantive emphasis]*
> – *and to do that via procedures P_1, P_2, ..., P_n [methodological emphasis]*

- *because of theoretical arguments T_1, T_2, ..., T_p*
- *and empirical arguments E_1, E_2, ..., E_q.* (p. 67)

According to van den Akker (2013), success in other contexts and in a framework of other projects increases the chance of knowledge growth, which justifies the consideration of other research projects with similar aims in section 2.1.5.

Here, the challenge is to design an intervention for supporting first-year engineering students in mathematics. The basis for the design of effective interventions is to know the context and the particularities, to understand what problems first-year engineering students face in mathematics, and what approaches to overcome these problems seem promising. These issues were addressed in Chapter 2: In summary, first-year engineering students face the problem that advanced mathematics at tertiary level is different from what they encountered at school. It demands assimilation of established concepts, abstraction from the concrete, and yet a flexible mind in order to reach an understanding deep enough to satisfy and to be applicable in new situations. The approaches that seem promising to overcome these problems address context, social aspects, learning resources, and learning strategies, as well as affective aspects like beliefs, attitudes, and approaches to learning. The theoretical arguments for this are given in detail in the corresponding sections in Chapter 2. The specification of these guidelines into tangible interventions and principles constitutes the focus of the following Chapter 4 on design development.

So, the main aim is to explore which characteristics of the intervention work in the given scenario, what procedures support them, and what (theoretical or empirical) arguments are found in order to understand the reasons. Therefore, the research questions are:

RQ1 What procedures can specifically support first-year engineering students in mathematics?

RQ2 What are the characteristics of an intervention for supporting first-year engineering students in mathematics?

RQ3 What combination of interventions is appropriate to promote learning strategies (as well as motivation)?

RQ4 Which learning behaviour is connected with academic success?

RQ5 How does MP^2-Math/Plus influence learning strategies (and motivation)?

Research questions one, two, and three concern the design of MP^2-Math/Plus, while research questions four and five address its impact.

The first research question aims at the practical realisation of the demands from section 2.1.2 (particularly page 42): Which learning resources should be covered during MP^2-Math/Plus tutorials? How can they be presented to the students in a way that encourages them to try out new modes of learning? What exactly promotes feelings of autonomy and self-efficacy in the learners? Which activities, methods, and attitudes reach out to them in a way that influences their motivation and learning?

The second research question takes a step back from the practical implementation and concerns the characteristics of the interventions. It covers not what can be done, but what are the criteria for choosing what should be done. Is it more advisable to help students in every possible respect, or (after initial introduction to new learning resources) leave them to fight on their own? How can support and independence be balanced? Which requests should be heeded, which should be denied? During the active project phase, is there a progression? If so, what are the guidelines?

The third research question regards the different elements of the intervention not separately, but as a whole. Are there interventions that advance each other? Which are good combinations? Are there maybe others that interfere with each other, or that are even counter-productive? It is also thinkable that some previously favoured interventions are expendable. This especially applies to those connected with relatively high expenses.

The next two research questions aim at the impact and evaluation of MP^2-Math/Plus. The effectiveness of the project is an important issue. It is the last of the four quality criteria[4] listed by Nieveen (1999). To answer this question, the group of project participants will have to be compared to non-participants. It is not clear if these groups are easily comparable because those participating will have applied for MP^2-Math/Plus, presumably with a need for support – a condition the other groups may not be able to fulfil. In Chapter 5, statistic test procedures will be employed in order to give a satisfying answer. As

---

[4] See page 85.

learning strategies (and motivation) are in the focus of the MP^2-Math/Plus interventions, it is of special interest to explore if and how they developed under project influence. The questionnaire in use registers different types of learning behaviour which can be considered both separately and in combination.

The following Chapter 4 will develop what characteristics and procedures in particular were decided upon in the first project cycle, and how they were re-designed in ensuing project cycles. Empirical (and other) arguments will be elaborated upon in Chapter 5 to back possible modifications.

4 Design Development

This chapter, together with the previous one explaining the research approach (Chapter 3) and the following one presenting empirical evidence (Chapter 5), forms a main part of this work, see orientation figure below.

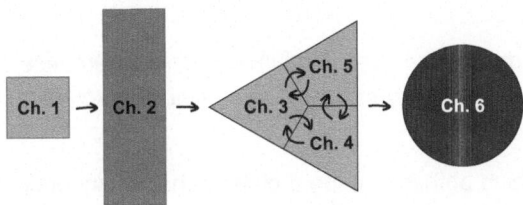

On the basis of the *Design Research* approach, this chapter provides an access (section 4.1) and gives an overview (section 4.2) of the conception of MP^2-Math/Plus, and of its development (section 4.5). It contains descriptions of the project procedures (section 4.3) for the different project cycles and gives reasons for the decisions to re-design some of them (section 4.4). In doing so, this chapter addresses the research questions on project design and development.

4.1 MP^2-Math/Plus Conception

MP^2-Math/Plus is one of the two parts of MP^2-Math/Plus/Practice which is based at Ruhr-Universität Bochum (RUB) and was one of six initiatives to win a prize in the competition *Nachhaltige Hochschulstrategien für mehr MINT-Absolventen*[1] in 2010 (Hetze, 2011). From 2010 to 2012, the project was

---

[1] Sustainable academic strategies for more graduates in mathematics, computer science, sciences and technology

sponsored by *Stifterverband für die deutsche Wissenschaft*[2] and *Heinz Nixdorf Stiftung*[3].

The original aim of MP^2-Math/Plus/Practice, according to its inventors, was to prevent avoidable dropping out of engineering courses (Dehling, Glasmachers, Härterich, & Hellermann, 2010). In other words, the project was not meant to look after those students who are either well capable of coping on their own, or who lack basic skills and interests necessary for a demanding university course.

MP^2-Math/Plus/Practice is based on two hypotheses:

1. Many students fail in mathematics because of their lack of self-organisation.

2. Many students lose their motivation for mathematics and willingness to work because they cannot see sufficient practical application of the subject material.

The idea behind both is that the problems that engineering students face in mathematics at the beginning of their studies are not primarily caused by the abstraction level of the mathematics they have to learn. As shown in section 2.1.6, the contents of mathematics in the first semester are in fact very close to, and in some parts identical with, school mathematics. The inventors of MP^2-Math/Plus/Practice looked for another reason for the modest pass rates. They settled for the idea that it is due to a lack of learning strategies, study skills and motivation that so many students drop out after the first semesters (Dehling, Glasmachers, & Härterich, 2012).

Therefore, MP^2-Math/Plus/Practice is structured in two parts, each addressing one of the basic hypotheses presented above. The first project part, Math/Plus, refers to learning strategies, the second part, Math/Practice, to mathematical applications relevant for prospective engineers and thus to motivation. With its project procedures, MP^2-Math/Plus is meant to support students' learning strategies by the example of mathematics, and hence strengthen them in other subjects as well because the impression that *Mathematics is what makes students drop out* can be reversed into *Who copes*

---

[2] German donators' association for sciences, see www.stifterverband.de.

[3] A donation founded by computer pioneer Heinz Nixdorf whose company was based in Paderborn, Germany. Nixdorf's company was acquired by Siemens in 1990, for more information see www.heinz-nixdorf-stiftung.de.

with mathematics copes with the rest[4]. Specifically, the project part Math/ Plus is intended to aim at supporting those engineering students who have realised they are facing substantial problems and are prepared to put in work and effort to overcome the obstacles. Therefore, some time at the beginning of the university course was needed to make them discern that substantial problems exist. Thus, MP²-Math/Plus does not start at day one at university, but allows students a few weeks to experience university life, i.e. to attend lectures, to go to tutorials, and to struggle with the weekly homework and the reading material. Despite the fact that information booklets about engineering stress the need for a substantial knowledge in mathematics, many students are confounded by the amount of work necessary for their mathematics lecture alone.

MP²-Math/Practice is intended to address well-performing students with motivation problems due to a lack of engineering applications. The advertising for this project part was therefore aimed at students who have passed the mathematics examination in the first semester. The capacity of MP²-Math/Practice (30) is considerably smaller than that of MP²-Math/Plus (120, counting groups that were supported in some way). The extra work associated with Math/Practice is credited with 3 points, and could only be expected to be accomplished by academically strong participants.

The two project parts with their distinct background hypotheses are separated by the university semesters: Math/Plus takes place in the first (winter) semester at university[5], when students are in the phase of having to adapt their learning behaviour from school routines to the new approaches that work at university. The perception was that, while they possess adequate motivation, at this stage students are not familiar with learning strategies that enable them to follow the methods and pace at university. The second project part, Math/Practice, aims at supporting students who are losing their motivation at the beginning of the second (summer) semester[6]. This project part offers them the possibility to work on several projects where the necessity of e.g.

---

[4] As mentioned in *Wer Mathe schafft, schafft auch den Rest*, see http://www.derwesten.de/staedte/ bochum/mathe-ist-der-knackpunkt-aimp-id7646982.html.

[5] Lectures in winter semesters in Germany last from October to mid February, with a two-week break over Christmas and the new year.

[6] Lectures in summer semesters in Germany start after Easter and go well into July, at RUB with a week's break over Whitsuntide.

differential equations becomes obvious. Math/Practice is described in more detail by (Härterich & Rooch, 2013).

4.2 Designing the First Project Cycle

In order to look upon MP^2-Math/Plus as a *Design Research* project, the structure by van den Akker (2013)[7] seems germane, listing the **function / purpose**, the **context**, the **characteristics**, the **procedures** (interventions) and **methods**, as well as the **theoretical** and **empirical arguments**. This set of statements will change or get new accents from one project cycle to the next. Whenever this structure is referred to, the key terms will be highlighted in bold face, like in this paragraph.

The onset for the first project cycle is the plan

- to design an intervention with the **purpose** to support first-year engineering students who both need and want support in mathematics,

- **characterised** by a substantive emphasis on learning strategies,

- **methodologically** via different support concepts (one with close and personal contact, one with remote and rather technical support, and a third control group without any support),

- which entail different **procedures** (see Figure 4.1),

- because of the **theoretical arguments** about learning behaviour presented in section 2.2.

This will allow to collect data in the first project cycle to compile **empirical arguments** for subsequent project cycles, particularly on the acceptance and efficiency of the procedures and characteristics.

In the initial project cycle, the project **procedures** comprised, as a starting point and a first attempt to answer research question one (see section 3.3),

- preparatory tutorials, where the regular tutorials were prepared with the help of more universally applicable learning strategies,

---

[7] See section 3.3.

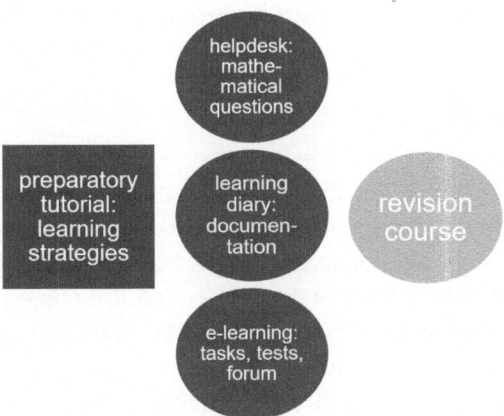

Figure 4.1: Procedures in the First Project Cycle, for SLG (blue / everything but revision course), SDG (ellipses), and MG (green / only revision course).

- a special MP^2-Math/Plus helpdesk where project participants could get their mathematical questions answered,

- a learning diary which would document learning time and style, and suggest e.g. planning learning sessions,

- an e-learning course with additional learning material, tests, and a discussion forum, and

- a revision course for the conclusive examination (see Figure 4.1).

All **procedures** were open to all MP^2-Math/Plus participants, with two exceptions: The preparatory tutorials were only for students in the innermost group with close and personal support (named SLG, *Supported Learning Group*); and the revision course was only offered to the other two groups (named SDG for *Self-Directed Group*, and MG for *Monitored Group*). Support contracts were made to bind participants from all three groups to attend regularly. More information on the **procedures** from the first project cycle is given in the next section 4.3, a retrospective assessment follows in section 4.4. In the WiGeMath framework (see section 2.1.3), MP^2-Math/Plus covers most of the objectives

(Table 2.7) and features the majority of possible impact and effect variables (Table 2.8).

4.3 Math/Plus Procedures

When the students had written the first mini exam (see section 2.1.6) and many accordingly made the experience that in spite of diligent preparations and a lot of effort, the mark did not satisfy their expectations, advertising for MP^2-Math/Plus started. The grading of the mini exams also offered the possibility to select those students who scored in the middle range, this way aiming at the individuals who fit the project **purpose**. A project flyer was stapled to the mini exams with points in the second lowest quarter, prior to being returned. The flyer introduced MP^2-Math/Plus and clearly stressed the fact that the defining **characteristic** was strategies for self-organised and efficient learning, e.g. independent follow-up of the lecture contents through temporary intensive support[8]. The MP^2-Math/Plus **procedures** (interventions) were outlined in the flyer as well.

In 2010, in the first cycle of MP^2-Math/Plus/Practice, 783 students appeared in the lecture statistics[9] (with similar numbers for subsequent years, and a slight increase after the double school leaver year in 2013). 618 of them actually took part in the first mini-exam, 317 of those scored in such a way as to be placed in the immediate target group (more than 4 and less than 9 points in the mini exam with a maximum of 12 points). The project was promoted especially to these students. To further ensure the **purpose** of selecting the desired target group, the results of the first mini exam in combination with an application which demanded reasons for wanting to take part determined if a student was admitted into one of the three central groups.

---

[8] „Für ein erfolgreiches Studium benötigst Du Strategien für selbstorganisiertes und effizientes Lernen. [...] So unterstützen wir Dich z.B. bei Deiner eigenständigen Nachbereitung des Vorlesungsstoffs durch zeitlich begrenzte intensivere Betreuung."

[9] This means that 783 students handed in homework, or took part in one of the mini exams, or in the final examination. The number of students registered for the courses in question exceeded 1,000.

4.3.1 Grouping of applicants

In the first project cycle in 2010 more than 200 applications were registered, some of them impressively stating the need for further help and in their urgency stressing the seriousness of the applicant, see section 5.1. The project **methodology** meant that out of those, 180 were selected and divided into three major groups: Three *Supported Learning Groups (SLG)* with 20 students each got the biggest amount of support, see Figure 4.1. Another 60 students were put into the *Self-Directed Group (SDG)*. The remaining 60 students formed the *Monitored Group (MG)*, planned as a control group. They did not get any specific help. This setting allowed to collect **empirical** data for separately evaluating the efficiency both of the preparatory tutorials and the e-learning course, in combination with a learning diary. Pseudo formal contracts were drafted that participants from the three groups were obliged to sign, together with their tutors from the preparatory tutorials (in case of SLG) or the project management (SDG and MG).

4.3.2 Preparatory tutorials (*Vorbereitungsübungen*)

The students organised in the SLG met once every week for 90 minutes with a tutor in a preparatory tutorial. In order not to mean further practise time (which the students already had in abundance, see section 2.1.6), the basic **characteristic** was to aid and to support, particularly in questions of self-organisation and structured learning, according to the design **principles** summarised in Chapter 3. Dealing with frustration and crises in motivation was not an aspect originally, but proved unavoidable.

The sessions of the preparatory tutorials were organised as follows, where the first three are described as an example. In the first session, the initial introductions were made, disclosing given names[10], place of residence, engineering course, and the favourite place for learning. After some organisational issues[11], a feeling of belonging and eye level contact was aimed at. In this context it was important to stress that MP²-Math/Plus is not connected to the chair of the professor holding the lecture, meaning that information revealed in

---

[10] The German custom to use formal address among adults is not strongly established among students and teaching assistants, but is standard between teacher and student.

[11] Such as access to e-learning, links for the online learning diaries or the difference between the university's official SZMA helpdesk and the MP²-Math/Plus HelpDesk.

confidence would not be carried further. A misery round (*What is your most urgent mathematical problem at the moment?*) led to a list of notions that posed problems. This list was immediately processed in jigsaw[12] phases, after the tutor had explained the notion nobody else dared to choose. The idea was to upload short explanatory texts and examples for each notion onto the e-learning course, so that efficiency and sustainability were guaranteed, as well as commitment to the group. The first session ended with the homework to daily fill in the learning diary and to mail a photo of the place of learning at home. The second session started with practising of names (after taking photos in the first session, the tutors learned the names, thus demonstrating the efficiency of using flashcards) and general inquiries into the state of mind. Then the photos of the participants had to be matched to the photos of their workplaces, inspiring a discussion on the advantages and disadvantages of the different layouts of desks and general work environment. This led to the compilation of a list of criteria for a good work environment, collected on cards. The cards were rated via sticky dots: Each student got five sticky dots and could decide how to distribute them among the cards. It was possible to stick all five on one card, or one each on five different cards, or any other thinkable way. Thus, the group result was a ranking list about how to organise your work environment. The next topic on how to structure the learning materials, followed naturally. In a think-pair-share[13], students collected their routines so far, discussed with a partner if that had proved useful at university, and shared their findings with the group. In the end, they were encouraged to make a resolution and to report back to their partners in the following week. The outlook on the third session was to consider reasons for and against notes on paper versus digital notes. The third session again started with practising names, this time placing the students under greater obligation. As some people's names are always easier to remember than others, this prompted a discussion on how to transfer this to the learning of mathematics, and on how to use several channels of perception

---

[12] Jigsaw (*Gruppenpuzzle*) is a method in two phases. First, expert groups (AAAA, BBBB, CCCC, DDDD) work on a task or problem with the aim of enabling every group member to be able to pass on the knowledge. In phase two, mixed groups (ABCD, ABCD, ABCD, ABCD) are assembled, where each groups member explains the results from the first phase to the others. The jigsaw method is communicative, efficient, and demands commitment.

[13] Think-pair-share prevents having a few dominate a group, and usually results in a wider range of ideas than brainstorming in plenum.

(Vester, 2004). These denote absorbing information in a way appropriate for mathematics, see section 2.1.2.

- auditively, by listening to it. In university mathematics, this means attending the lectures, paying attention and discussing the subject matter with peers;

- visually, by using graphics and pictures. To make the most out of this channel, the use of colours, movement (dynamic software), and space is recommended;

- haptically, by touching (or writing, drawing, designing etc.); and

- cognitively, by connecting new subject matter to established knowledge (indispensable in mathematics).

Furthermore, it can help remember mathematics by becoming active (instead of passively copying notes and solutions), and by enlisting positive feelings (e.g. meeting new friends for a learning session in mathematics). The options, possibilities, and estimations were discussed in detail during the third group session.

The subsequent sessions were organised in a similar fashion. For example, once the focus was on time management. This included scheduling a week, including recreation, sport, and social contacts. The time needed for work on the weekly assignments was planned, as first-years seldom manage to solve all the tasks in one go. Time management also refers to planning the rest of the semester: compiling an overview over what has been done and what still needs to be done, and allowing realistic time for the steps in between. Another session was dedicated to getting help from computer algebra systems, in this case from GeoGebra[14] and Sage[15]. Other sessions concentrated on sifting through books that might be helpful for different purposes, testing websites relevant for engineering mathematics, or trying various ways of memorising mathematical facts and formulae.

---

[14] geogebra.org
[15] sagemath.org

4.3.3 MP^2-Math/Plus HelpDesk

Despite the helpdesk offered by RUB through its mathematics service centre SZMA (see section 2.1.6), the students participating in MP^2-Math/Plus were in need of additional help in mathematics, in accordance with the design **characteristics** based on the insights gained from von Glasersfeld (1991), see Chapter 3. A special MP^2-Math/Plus helpdesk was introduced to remedy this issue. Three senior students of mathematics were hired for altogether twelve hours a week. They were instructed to answer questions, to help find mistakes in calculations - and to refrain from lecturing or from merely presenting correct solutions. SLG and SDG students were invited to use this **procedure** to ask their questions about the weekly assignments or the lecture itself.

4.3.4 Learning diary

The **purpose** behind the learning diary which was used by SLG and SDG was twofold: First, it was meant to remind students of successful learning strategies by daily suggestions, e.g. learning with fellow students or planning learning sessions, in order to provide alternative learning scenarios (see Chapter 3). What is more, there were questions about motivation, mental state and retro-spectively about contentedness with the learning session. The **purpose** was that students would reflect on and modify their learning behaviour by using the learning diaries. In addition to that, a learning diary was regarded as the only practical means to really investigate students' learning behaviour. As the bulk of study work has to be done out of lectures and tutorials (see section 2.1.6), there is no other way than to ask the students in some way about their learning behaviour. The learning diaries were to yield detailed information on where students studied, with whom, for how long, which strategies they used etc., so that this would in turn reflect in the results they achieved in the examinations, or at least in the attitudes towards learning in general or mathematics in particular, as conceptualised in Chapter 3. The learning diary introduced in the first project cycle in 2010/2011 was based on Landmann and Schmitz (2007a) and can be found in Appendix D.

4.3.5 e-Learning course

The MP^2-Math/Plus e-learning course was the central **procedure** to contain uploads explaining mathematics, links to related sites, tasks and tests to assert one's abilities in mathematics, and a forum for discussion and exchange. The course was housed in *moodle*, one of the university's e-learning platforms. Some effort was taken in order to assemble tests with automatic evaluation to make sure the students had enough practise material, in compliance with the demand to present the subject matter in context (Dreyfus, 2012) and treat the students as independent learners (see Chapter 3). The lecture script was uploaded, as well as information on the organisation of the project, e.g. times and topics for the repetition course and the MP^2-HelpDesk.

4.3.6 Revision course (*Repetitorium*)

A concise and compact revision is a good idea in the finish to an important examination, but this was not the only reason that a revision course was viewed as a useful **procedure** for MP^2-Math/Plus. Another point was that it was thought to serve as an incentive for students in SDG and MG to regularly fill in the learning log, as that was suspected from the beginning to seem rather tedious. The **purpose** of the revision course in the first project cycle was to confront students with typical examination tasks and to work out typical solutions. The selection of what was typical was based on examination papers from the years before which were available through the student councils. The date for the revision course was set after lectures had terminated, but well before the written examination in order to allow students time for catching up in case the revision course revealed too many or too large gaps.

4.4 The First Project Cycle in Retrospect

It is remarkable that the majority (82%, see section 5.1) of students applying for MP^2-Math/Plus did so because they wanted to overcome the *mathematical* problems they had experienced in the first weeks at university, when the advertising and presentation of the project clearly stressed the **characteristic** learning strategies. Participants obviously conceived the subject-specific obstacles as overwhelming. When the kind of problems were described, they ranged quite equally from understanding and transfer (32%), lack in basic skills (28%), speed and routine (25%) to learning behaviour (25%) and specific knowledge gaps (23%), see section 5.1. The level of suffering seemed considerable, as some verbatim utterings showed, e.g. "I hope to finally keep up in mathematics and to not always lag behind because I don't understand a thing, no matter how much I try to think my way into it, mathematics is one of the subjects in this university course where nothing seems to work out"[16].

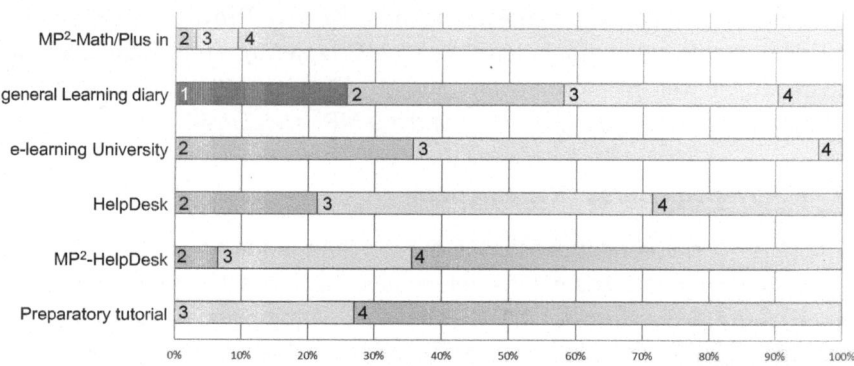

Figure 4.2: First Project Cycle, *Would you recommend the intervention?*
Likert scale from 1 (not at all, red) to 4 (very much, green), $n = 38$.

As a first step to evaluate the MP^2-Math/Plus design after completing the first project cycle, participants were asked to fill in a questionnaire (see Appendix C) where they could give their opinion anonymously. The answers to the questions if they would recommend the individual project interventions are presented in Figure 4.2: The vast majority of students would indeed recommend MP^2-Math/Plus as a whole, even more than they would recommend any of the single **procedures**, so the sum of the project is rated better than its parts. The individual project interventions are arranged in Figure 4.2 from less recommendable to highly recommendable, and it is striking that the learning diary scores lowest, it is the only one with any ratings of *not at all recommendable*, and these amount to about 25%. This clearly is an assignment to

---

[16] Originally „Ich verspreche mir endlich mal in Mathematik mitzukommen und nicht immer hinterher zuhängen, weil ich nichts verstehe egal wie sehr man sich versucht reinzudenken, Mathematik ist auch mit eines der Fächer in diesem Studiengang wo nichts klappen will".

re-design this **procedure**. The e-learning course received the second lowest recommendations, and indeed the performance here had not met expectations: Apart from technical problems, the tests (created by a student assistant) did not convince, the forum was left abandoned, and the few uploads did not keep the promise of being particularly helpful. It may be that students did not want to log onto yet another platform (apart from *Blackboard*, where the mathematics lecture was housed, and *Flexnow* which held the recognition of credits, and probably social networks like *facebook* where they communicated with their friends), or that they simply did not feel the need to create a new collection of explanatory material, when the Internet was already full of it to a much greater extent. The comments in the evaluation survey support this perspective: "Pity that only a few joined in.", "A good idea, unfortunately not honoured by the students"[17].

The university's SZMA helpdesk was assessed decidedly lower than the MP^2-Math/Plus HelpDesk. This can be explained by the circumstance that student assistants are obliged to hold their office hours in the university's SZMA helpdesk, which is often regarded as an additional duty to the teaching and marking they are officially being paid for. The SZMA helpdesk is open to all students who attend courses in service mathematics, i.e. students of engineering, biology, chemistry, geology, psychology etc., which represents a huge demand on the staff, or, as one student put it in the evaluation survey, "it seemed to me as if the staff there did not really feel like explaining anything"[18]. On the other hand, the MP^2-Math/Plus HelpDesk displayed different **characteristics**: The student assistants there were explicitly paid for manning the helpdesk. They supported the students to find solutions themselves and were expected to only cater for the one specific mathematics lecture involved in MP^2-Math/Plus. There were always the same staff to be found, so it was possible to take up a problem from the week before. This is illustrated by several of the comments in the evaluation survey, e.g. "I got help on every topic and with this help was able to solve all weekly assignments" or "The girls were awesome, competent

---

[17] Originally „Schade, dass sich dort nur wenige beteiligt haben."or „Eine gute Idee, die leider von den Studierenden kaum angenommen wurde."
[18] Originally „Ich war nur einmal da, aber es kam mir so vor, als hätten die Mitarbeiter dort selbst nicht allzu viel Lust, das zu erklären."

and friendly"[19]. The room originally made available for the MP^2-HelpDesk often proved too small, and the students had to use a bigger nearby seminar room with more seating.

The key **procedure** of the project, the preparatory tutorials, received the highest recommendation. Nevertheless, there was criticism expressed in the evaluation survey: Several students wished for less focus on learning strategies and more tuition-like demonstrations of exercises. So short before the final examination, the need for even more subject-specific support is understandable, but it stands in opposition to the MP^2-Math/Plus **characteristic** of aiming at learning strategies and of *not* substituting the regular lecture or tutorials, and was therefore not included as a perspective for re-design. This feedback had an influence on the decision to open the revision course, another **procedure**, to the members of SLG, too, as it would meet this need. Other comments on the preparatory tutorials were full of thanks and appreciations, e.g. "I felt in good hands and respected with my personal mathematical problems" or "A very helpful programme to smooth the start at university"[20].

Apart from the students' assessments illustrated above, the main **purpose** of MP^2-Math/Plus is to help students pass the written examination in mathematics. So the examination statistics (see also section 5.5) are the key **empirical argument** on which to test the success of the project. They can be found in Table 4.1. These statistics do not take into account that the students addressed by MP^2-Math/Plus deliberately come from the lower (though not the lowest) performance spectrum, meaning that the numbers for non-participants also include excellent students. Their pass rate of 61.17% (calculated in reference to those attending the examination) is not surprising. The fact that there is statistically no difference between the three groups SLG, SDG and MG is disappointing. In absolute numbers, SLG, SDG, and MG are on the same level with 20, 19, and 18 passes. Students from SLG attended the examination in a slightly higher proportion (68.33%) than SDG and MG (58.62% and 63.64%,

---

[19] Originally „Es konnte mir bei jedem Thema geholfen werden und mit dieser Hilfe konnte ich alle wöchentlichen Übungsaufgaben lösen", „Die Mädels waren echt super, kompetent und freundlich."

[20] Originally „Ich fühlte mich gut aufgehoben und auch mit meinen persönlichen mathematischen Problemen respektiert.", „Ein absolut hilfreiches Programm, um den Einstieg in die Uni zu erleichtern!"

respectively). This can be interpreted as a success, as they felt obliged not to give up prematurely, but it leans on the pass rate.

Table 4.1: Examination Statistics for the First Project Cycle

Project Group	Total	Attended Exam (%)	Pass (%)
SLG	60	41 (68.33%)	20 (48.78%)
SDG	58	34 (58.62%)	19 (55.88%)
MG	55	35 (63.64%)	18 (51.43%)
non-MP^2	883	649 (73.50%)	397 (61.17%)

Another issue influencing the re-design for the second project cycle were comments by participants who in retrospective declared that they should have worked more over the course of the complete semester than they had actually done. Looking back, they agreed that it would have been a good idea to really devote themselves to the weekly assignments. This was unanticipated, as to promote continual work from the beginning was one of the **characteristics** of MP^2-Math/Plus, and had been a focus of more than one of the preparatory tutorial sessions. It seems that hearing good advice does not necessarily equal acting on it, a point leading to a search for **theoretical arguments** remedying this problem.

4.5 Consequences and Re-Design

The **empirical findings** (particularly the fact that students who worked more independently were as successful as those under close care) and further **theoretical arguments** led to the following adaptations of the project design, according to the structure by van den Akker (2013):

In order to realise the emphasis on learning strategies, the procedures had to be **characterised** by additionally addressing motivation and affective aspects, too, see research question two (section 3.3). The fundamental conclusion from the first project cycle was that the presentation of alternatives alone may not entail a change in learning behaviour. Students are more inclined to follow advice when their affective needs are met. In order to keep up their motivation to change their learning behaviour, the **theory-grounded** recommendation by Deci and Ryan (2000) is to consider competence, autonomy, and relatedness. This was assisted by the mentors (see section 4.5.3) who would provide an outlook on one's own situation a year later and also fits Heckhausen (1977) and colleagues who emphasise recalling the possible outcomes of the actions taken (see section 2.1.2). So the additional **purposes** were to exact commitment and self-reliance from the students, in exchange for relatedness, autonomy and competence.

To implement relatedness, the **characteristic** role of the MP^2-Math/Plus groups and their tutor was strengthened, among other things by introducing social networks and a support agreement in written form[21]. The team spirit was to be promoted by suitable activities, including the mentors and group tutors. The groups meetings ("preparatory tutorials" in the first project cycle) were re-named "learning groups". This symbolises the intention to shift the responsibility for learning success from the tutors to the students who were to become active agents of their learning.

Methodologically a progress towards independence had to be incorporated into the project conception. The project time was therefore divided into three almost equally long phases of four to five weeks each. The first phase lasted from the beginning of the project until the Christmas break, the second until the end of lectures in mid-February, and the third and last until the written examination at the beginning of March. In the first phase, the students got every help and support. In the second phase, their questions were answered by asking back *(Where can you find this information?* or *What can you do to solve this problem?)*. In the first two phases, students were more and more encouraged to engage in peer work, and to help each other. In the third phase, the group tutors refrained from structuring the group meetings. This strategy makes sense in the light of the fact that MP^2-Math/Plus would not be there to help after the first semester – but their peers and other resources would.

Thus, in the course of the semester a progression was planned from more supportive to more independent work. This also implied that students could be expelled if they did not keep to the rules (e.g. if they did not attend the MP^2-Math/Plus HelpDesk or the meetings or if they did not fill in the questionnaires). That is why groups started with 25 students (five more than in the first project cycle). Around Christmas a first conclusion was drawn about who was

---

[21] The support agreement replaced the support contract from the first project cycle. It stressed the eye-level position by listing as many obligations for the project participant as for the project tutor, and needed signatures from both, see Appendix E.

allowed to stay. These regulations were communicated to the students from the beginning, so that they were prepared. As a result a few students were actually expelled, though none against their will.

The project **procedures** were modified as a consequence, resulting in a refined answer to research questions one and two (section 3.3), see also Chapter 6.

4.5.1 Learning groups

In the second project cycle in 2011/2012, the weekly meetings of the *Learning Groups* took place on Fridays and were divided into the three phases described above. To promote team spirit and facilitate debate[22], the seating arrangement was changed into more communicative group tables, see Figure 4.3.

Figure 4.3: *Learning Group* Seating Arrangement, photo by Michael Kallweit

Phase 1 (4 meetings) The students got information on the project and its activities. They got to know each other, their tutor, the students working in the MP^2-Math/Plus HelpDesk and their mentors. They reflected the main **characteristic** of the project, learning strategies, particularly the learning strategies they

---

[22] (Tall, 1991b) sees debate as essentail for the construction of meaningful concepts, see section 2.1.2.

had used so far. They learned facts worth knowing about learning and working strategies and put them to the test, e.g. time management, (self-)motivation and taking notes. Another important point was the constant encouragement to seek help and support from the MP^2-Math/Plus HelpDesk, from the SZMA helpdesk, from fellow students, from tutors, or from those working in MP^2-Math/Plus.

In the last meeting before Christmas (Dec 23rd, 2012) GeoGebra and WolframAlpha were presented, as they could offer support particularly during the holidays when the campus was not accessible.

Phase 2 (4 meetings) This phase can be summarised by looking back and ahead: on the (almost) past semester and on the time left till the written examination. Now the **purpose** was to structure the remaining time in order to master the content knowledge in a feasible pace. Beside the obligatory **characteristic** of time management, there were exercises from past examinations, and a personal strategy for the examination was to be devised. The collaborative search for mistakes in fake solutions enhanced the awareness for typical mistakes and encouraged intense preoccupation with exercises. Of course, more than before, the focus was on subject-specific questions, as the final examination was drawing near. It was stressed that explanations did not have to come from the tutor, but could be learned from other sources as well, e.g. such as online tools, peer students or books.

Phase 3 (5 meetings) The students worked on exercises from the homework assignments, from past examination papers and from books. Whenever they applied to the tutor for help, the **method** was to redirect them to their fellow students or to summaries worked on in previous sessions. The tutor now only served as a brace to mark the beginning and ending of the group meetings; many students continued learning together after the official time was up. The *Revision Course* took place simultaneously.

In the very last week before the written test, which was fixed for the March 12, 2012, neither *Revision Course* nor group meetings took place. Nevertheless the MP^2-Math/Plus *HelpDesk* was open. The idea behind was sensible time management – it is not advisable to work on new material until right before the examination.

4.5.2 Linking project procedures

Another **characteristic** came from the circumstance that the individual project procedures had existed rather independently in the first cycle. Thus a closer connection and reference among them was aimed at by addressing the same aspect from various directions through the different interconnections between the project **procedures** (see Figure 4.4). This means that from the second project cycle on, there was a weekly focus topic that was worked on in the group meetings, in the newly introduced workbooks and in the e-learning course.

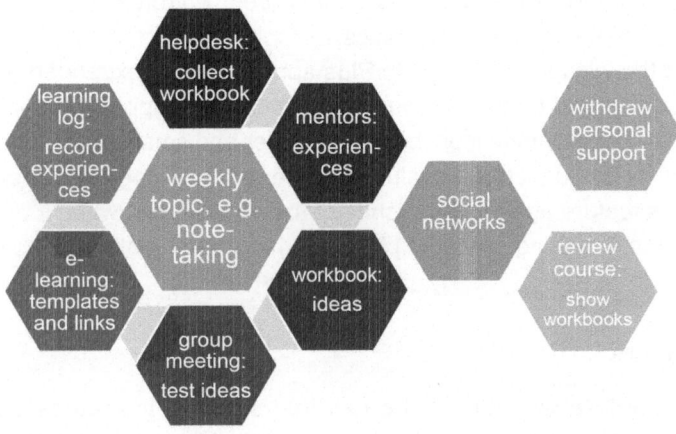

Figure 4.4: Linking Project Procedures in the Second Project Cycle, Note-taking as Example

One example for linking the procedures is note-taking, it is one of the topics chosen for every project week. The participants had to attend the MP²-Math/Plus *HelpDesk* (see section 4.5.10) in order to collect the weekly *Workbook* (see section 4.5.8) which contained statements by the *Mentor* students (see section 4.5.3) from the year before about their experiences with note-taking, and some more ideas and tips on note-taking. In the group meeting (see section 4.5.1) that week, the students were asked to try out some of these ideas, and when they logged on to the e-learning platform (see section 4.5.5), they found useful links on that topic, e.g. to webpages where you can create your own, individually structured notepaper. In the *LearningLogs* (see section 4.5.4), which as weekly versions replaced the daily learning diaries, the students recorded their experiences with the different types of note-taking. Social networks like facebook (see section 4.5.6) were used to start a post to invite students to share their experiences on note-taking. And in order to attend the *Revision Course* (see section 4.5.9) shortly before the written examination, the students had to show all the weekly *Workbooks*. This strong network made another **characteristic** possible, namely withdrawing immediate support, so that students could become more and more independent, and the project tutors became more and more indispensable. All this was also meant to lead to more obligation to utilise all project procedures, to truly commit to studying engineering, respectively mathematics.

Due to the wishes of MP^2-Math/Plus applicants and expanded resources, the SDG, the group that had access to all technical support procedures, but did not attend weekly meetings, became smaller and smaller, which annihilated the possibility for quantitative evaluation. The **methodological** decision to not enforce this feature against the wishes of the participants, when it would additionally mean foregoing available places in the group meetings, was therefore taken.

4.5.3 Mentors

A new **procedure** was due to the fact that some of the advice given was not implemented. It entailed the idea of finding *Mentors*, students who had participated in the first project cycle and who could provide first-hand information that might be heeded to a greater degree. The intention was that students would feel closer to the *Mentors* than they could to the project tutors, who were doctoral students with degrees in mathematics. Meeting the mentors would also, in compliance to the recommendations (see section 2.1.2) by Heckhausen (1977), provide an outlook on how the students themselves might be feeling a year in the future.[23] Participants from the year before could support freshmen in other aspects, too, e.g. in relation to perseverance, examination preparation or dealing with frustration. And they could promote the project accordingly, thus demonstrating the self-efficacy important for keeping up motivation.

The criteria for choosing the *Mentors* differed from those for the employees in the MP^2-Math/Plus *HelpDesk*. For *Mentors*, it was regarded more important

---

[23] The transformation from insecure first-year to confident second-year was striking in some cases.

that they were easy to identify with, in accordance with the **characteristic** of relatedness, according to Deci and Ryan (2000), see section 2.1.2. Their mastery of mathematics was not so important, nor was the fact whether they had passed the mathematics examination critical. Nevertheless the circumstance that MP^2-Math/Plus had helped the prospective *Mentors* to come to terms with their learning behaviour, their time management and their motivation was essential. This led to the commitment of Maria, Marius and Lars. Maria had started UTRM in the year before, had worked very diligently and passed the mathematics examination with a good mark at the first attempt. Marius had enrolled in the BI course and had often been at odds with the complicated language used in mathematics lectures. Lars is a student of MB who had faced problems adapting to university life and had occasionally had problems to motivate himself.

The *Mentors* were asked to fill in a few questions (see Appendix F) on their learning behaviour to be used in the weekly *Workbooks*, see 4.5.8. Thus their comments could be used to illustrate their perspective on the weekly topics. The *Workbooks* also contained photos of the *Mentors*, and they visited the weekly *Learning Groups* two or three times during the semester.

4.5.4 LearningLog

A re-designed **procedure** referred to the conjuncture that the learning diary had neither met with acceptance (see section 5.3) nor had it supplied data expedient for insights into the consequences of different learning behaviour, see Griese and Kallweit (2014). This led to a reduction in quantity, a weekly (instead of daily) *LearningLog*. The intention of promoting the **purpose** to suggest new learning strategies and to encourage reflection was not realised. Therefore the *Learning Diary* was converted into a weekly form which contained the same group of questions about the past week, about the current state of mind and about the plans for the subsequent week. As before, it was possible to fill in an online (see Figure 4.5[24]) or a paper version.

---

[24] Translation of the 4-point Likert scale items (not true at all – exactly true): How do you rate your learning success from the past week? I am content with the relation between learning time spent in concentration to learning time altogether. I am content with the result of my learning. I overestimated my capacity today. I worked qualitatively well today. Today I reached my personal learning goals set before learning.

Wie beurteilen Sie Ihren Lernerfolg der vergangenen Woche?

3.5	Ich bin mit dem Verhältnis konzentriert verbrachter Lernzeit zur gesamten Lernzeit zufrieden.	trifft überhaupt nicht zu ○ ○ ○ ○ trifft genau zu	
3.6	Ich bin mit meinem Lernergebnis zufrieden.	trifft überhaupt nicht zu ○ ○ ○ ○ trifft genau zu	
3.7	Ich habe meine Arbeitskapazität überschätzt.	trifft überhaupt nicht zu ○ ○ ○ ○ trifft genau zu	
3.8	Ich habe qualitativ gut gearbeitet.	trifft überhaupt nicht zu ○ ○ ○ ○ trifft genau zu	
3.9	Ich habe meine persönlichen Lernziele, die ich vor dem Lernen festgelegt habe, erreicht.	trifft überhaupt nicht zu ○ ○ ○ ○ trifft genau zu	

Figure 4.5: Excerpt from the *LearningLog* Used in the Second Project Cycle

4.5.5 E-learning 2.0

Also, the **procedure** of the e-learning course had not fulfilled its potential, further effort was needed. The fact that the students had not communicated via the forum led to the decision to comply with their habits and include social networks in MP^2-Math/Plus, see section 4.5.6. The e-learning course in the second project cycle contained again tests, tips, links, uploaded commented solutions, and examination papers from the past. It was housed in moodle, as before. Its special highlight was the original hand-written lecture script especially processed as a pdf-file with a table of contents in a printer-friendly version that did not lose anything when hole-punched. The e-learning course was also used to announce the dates and topics for the *Revision Course*, see 4.5.9.

4.5.6 Social networks

The feedback that the students had refused logging into moodle inspired the **procedure** of meeting them halfway in the social networks they use on a daily basis for their private contacts. In order not to settle unilaterally for one of the social media platforms popular at the time, groups were started on Facebook, Twitter and Google+. This policy proved effective. The project participants communicated with each other and the tutors about exercises (like eigenvectors, see Figure 4.6) and organisational issues, e.g. if it made sense to attend the

Figure 4.6: Facebook Conversation from the Second Project Cycle

post-examination review. This stressed the feeling of relatedness, see Deci and Ryan (2000), section 2.1.2.

4.5.7 Mock test (*Probeklausur*)

Furthermore, a *Mock Test* (see section 4.5.7) was introduced to the range of project **procedures**. It was intended to confront participants with possible outcomes of different learning behaviour (as covered by the term competence, see Deci and Ryan (2000), section 2.1.2), and prepare them for the as yet unknown task type of multiple choice exercises, which constituted a substantial part of the written examination that year. As there is a great variety of possi-

Figure 4.7: Students Writing the *Mock Test* During the Second Project Cycle,
photo by Michael Kallweit

bilities for multiple (or single) choice tasks and of how to grade them[25], this
unknown situation caused uneasiness. It is a challenge to really experience
an examination under realistic conditions, in a separate room, at a designated
time, without even the theoretical help of a computer or a calculator, and under
supervision, see Figure 4.7.

The *Mock Test* assembled as a **procedure** for MP^2-Math/Plus contained
exercises on all the areas considered relevant for the examination. For the
reasons stated above, it was decided to design all eleven exercises as multiple
choice items, choosing formats that test both understanding and calculation
skills. In detail, the test items covered the contents given in Table 4.2.

Table 4.2: MP^2-Math/Plus *Mock Test*

Task	Contents
1	Linear dependency
2	Relative position of a straight line and a plane
3	Multiplication of matrices
4	System of linear equations
5	Convergence of sequences
6	Continuity of a function
7	Local minima and maxima
8	Derivative
9	Partial integration
10	Integrating by substitution
11	Linear differential equation, order 1

4.5.8 Workbooks (*Arbeitsbücher*)

Weekly *Workbooks* were another **procedure** in order to reach not only coherence and structure, but also relatedness (the *Workbooks* are exclusively for MP^2-Math/Plus participants), competence (via practical tips on how to improve your learning, given by the personally known mentors) and autonomy (they contained pages to be filled in individually, for individual choices), see Deci and Ryan (2000), and section 2.1.2. The *Workbooks* are little booklets that are collected in a blue folder with an MP^2 logo, see Figure 4.8. They contain some recurring categories (see Table 4.3, e.g. tips / statements from the mentors, wishing well and suggestion box, preview on the following week as well as spaces to fill in) - typical of the additional **purposes**. The general tone is casual and conversational. The *Workbook* titles are motor sports metaphors (start-up aid, navigational aid, upshift, training victory, pole position, pit stop, cruise control, home stretch, final spurt), see Appendix G.

---

[25] There is a huge difference, for example, if the number of correct answers is known or if wrong markings produce a deduction in the score (formula scoring). Also the number of options and the quality of the distractors weigh heavily on the validity of a multiple choice test. For more details on test construction and evaluation see Bühner (2011) or Rost (1996).

Table 4.3: MP^2-Math/Plus *Workbooks* in the Second Project Cycle

Recurring Categories	Weekly Topics
tips / statements from the mentors	1: introduction of people involved in MP^2-Math/Plus; project interventions and activities; contact information; seven truths according to (Zucker, 1996)
wishing well / suggestion box	2: MP^2-Math/Plus *HelpDesk*; more information on project activities, e.g. e-learning and social networks; motivation tips and learning techniques; overview on
preview on the following week	mathematics learned so far
	3: tips on time management; weekly plan; semester plan; access to learning; places for learning; learning
spaces to fill in	targets as motivational aid
	4: taking structured notes; learner types; learning with file cards
	5: planning non-term time; digital help (CAS, GeoGebra, WolframAlpha)
	6: intermediate conclusion; overview of the subject matter (basic formulae and rules, linear algebra, analysis etc.)
	7: information on non-term time (dates for *Revision Course*, MP^2-Math/Plus *HelpDesk*, *Learning Group*); work on old examination papers
	8: mathematics on paper (tips for writing down solutions); avoiding mistakes; typical mistakes
	9: preparing for the written test; to-do list; weekly timetable till the written test; test policy; interview with Nadine (tutor from last year), motivation list of treats / rewards for the time after the examination

Figure 4.8: *Workbooks* Used in the Second Project Cycle,
          photo by Michael Kallweit

4.5.9 Revision course (*Repetitorium*)

In the second project cycle, the *Revision Course*, the last **procedure** to be experienced before the end of the project, was open to all students from SLG and SDG (i.e. not restricted to SDG as it was in the year before), if they fulfilled the requirements, meaning they had filled in the *Learning Logs*, had visited the MP²-Math/Plus *HelpDesk*, and if members of SLG had regularly attended the weekly meetings of the *Learning Groups*. The *Revision Course* took place for two groups twice a week respectively, for three weeks. Each session lasted 90 minutes, and a small lecture hall seating up to 70 students was chosen for the purpose. The topics of each session were published in advance (true to the **purpose** of promoting autonomy, see section 2.1.2), so that students could decide if they wanted to attend a particular session. Together with the topics themselves, there was reading material from the published typed script, the handwritten manuscript and the textbooks recommended for the lecture. To round up the preparations, a limited number of exercises from Papula (2010,

2011, 2012) was suggested for each session (referring to the **purpose** of fostering competence).

Figure 4.9: The *Revision Course* During the Second Project Cycle, photo by Michael Kallweit

One example for a task from the revision course is the calculation of the indefinite integral

$$\int x^2 \cdot \exp(-x) \, dx \, .$$

The solution[26] was presented step-by-step, with auxiliary calculations for the antiderivatives of the factors for the integration by parts. As it is necessary to apply this rule twice, students were warned against losing a minus sign, the use of brackets was recommended and demonstrated. A check completed the calculations, a technique which can provide additional security in an examination. An alternative approach was presented, namely to start with the antiderivative

$$F(x) = \exp(-x) \cdot (ax^2 + bx + c)$$

with $a, b, c \in \mathbb{R}$. By deriving F and comparing coefficients, the same solution can be reached – without the need for integration in parts, simply by applying

---

[26] The indefinite integral is $\exp(-x) \cdot (-x^2 - 2x - 2) + c$ with $c \in \mathbb{R}$.

Table 4.4: MP^2-Math/Plus *Revision Course*

Session	Topics
1	sequences (bounded, monotonically increasing / decreasing, geometric sequence, Fibonacci sequence, recursive sequence, convergent with limit g, de l'Hopital, Cauchy sequence)
	series (partial sums, geometric series, ratio test, exponential function)
	continuity (definition, visualisation)
2	derivation (rules of derivation, in particular chain rule)
	optimisation exercises (objective function, side condition, testing of marginal values)
3	straight lines, planes (parameter notation, coordinate and normal form, Hesse normal form, interceptions and relative positions)
	vector algebra (scalar and vector product, angles, volume of a parallelepiped, area of a parallelogram)
4	matrices (determinants, inverse matrix)
	systems of linear equations (solvable / not solvable, homogeneous and inhomogeneous, solution space)
	eigenvalues and eigenvectors, transformation matrices
5	integration (partial, substitution, partial fraction decomposition)
6	differential equations
	parameterised curves (sketch, curve length, solid of rotation)

the product rule. This stresses the connection between the product rule and integration in parts, and anticipates the comparison of coefficients used in the solution of integrals via partial fraction decomposition.

4.5.10 MP^2-Math/Plus HelpDesk revisited

As the MP^2-Math/Plus-HelpDesk had been popular in year one (see section 5.5), not many **characteristics** were changed in the second project cycle. It was considered crucial to employ senior students well-qualified to work there. The criteria for choosing staff were the ability to emphasise with mathematical problems, enthusiasm for MP^2-Math/Plus and the idea of supporting engineering students, and didactic or methodical experiences in mathematics. Luckily three students were found who perfectly matched these requirements: Hannah, a *Master of Education* student of mathematics and biology who had before studied engineering; Jonas, a student of mathematics who had been a marking homework for engineering students in the year before; and Marius, an engineering student from the year before who had excelled in mathematics. Among them, these three covered different aspects of gender, university course, experience and proximity to the first year students.

4.6 Further Project Cycles

After major adaptations in the second project cycle, the third and later cycles underwent only smaller changes. These are presented in this section, as before in reference to van den Akker (2013) and constitute further response to research questions one and two (section 3.3), see also Chapter 6.

The evaluation of examination results from the second project cycle showed gender differences (see section 5.5.1): Female project participants profited from the project interventions to a higher degree than their male counterparts. This seemed surprising, as MP^2-Math/Plus was not conceptualised as a decidedly female-oriented project. This **empirical argument** implied further adaptations. Literature was studied in view of how to reach the **purpose** of appealing to males, of motivating them to live up to their potential (and not disregarding the females), leading to **theoretical arguments** about positive role models and attribution theories, see Lazarus (1991) and Weiner (1994). *Mentors* and *HelpDesk* staff were selected carefully, with regard to diverse ethnic, educational and gender backgrounds. The idea to appeal to male (aspiring) engineers via gaming elements was prosecuted by introducing new versions of the original learning diary, see sections 4.6.1 and 4.6.2 for more details.

The third and further project cycles also saw some changes in staff, in part due to the circumstance that MP^2-Math/Plus was extended to other engineering courses at RUB: *Electrical Engineering*, *Information Technology*, and *IT Security*. This was made possible by funding from the Reinhard-Frank-Stiftung[27]. The expansion also boosted the number of applicants to almost 200 and thus the number of learning groups up to eight (in 2015 / 2016). As the success of the second cycle induced the dean's office to supply more staff, it was

---

[27] http://www.reinhardfrank-stiftung.org/

possible to establish four instead of three *Learning Groups*. This practically annihilated[28] the SDG, as all students could be offered places in the *Learning Groups*, modulo their personal schedules. In collaboration with a sister project for bachelor students of mathematics, the advertising and internet presence was relaunched[29].

The *Workbooks* were adapted in reference to new *Mentors* and staff and now sported the university's corporate design in colour and an imprint. Due to slightly different times and dates, there were only eight instead of nine *Workbooks*. The *Mock Test* was also adapted as the mathematics lecture had a clearer priority on matrices, eigenvalues and eigenvectors, and differential equations had only just been introduced. The digital procedures were constantly amended, see sections 4.6.1, 4.6.2 and 4.6.3.

The original idea to wait until the results of the first mini exam did not stand up to practise, as in some years lecturers had terminated their mini exam at times inconvenient (too late) for MP^2-Math/Plus. But even relatively early mini exams had delayed the project start until the first days of December, which meant losing a number of weeks. As a consequence, the idea of waiting for the first mini exam feedback as a start signal for the MP^2-Math/Plus was abandoned, and the project start was brought forward to the end of October, only a few weeks after the lectures had started. The intention of using the mini exam results to acquire the intended target group for MP^2-Math/Plus, namely students who needed support in order to succeed at university, but who did not lack very basic mathematical skills was therefore lost. Retrospect analysis of the project group investigated the question if this led to more high-performing students participating (which may constitute an advantage for the learning groups, meaning more options for peer support), or if more academically weak students took part (which would impair pass rates, and complicate efficient subject-specific work), see section 5.1. The answers were inconclusive, so the early start was retained.

At the current day, Math/Plus is a well-established project at RUB, and is set to continue in future, offering support to engineering first-years, as well as bachelor students of mathematics.

---

[28] Only three students remained.
[29] rub.de/matheplus

4.6.1 LearningLog online

The project **procedure** concerning the digital support was re-designed more than once: In the third project cycle, an online learning log was introduced.

Although the documentation of learning behaviour had been reduced from the first project cycle to the second to a shortened and weekly form, the *LearningLog* (see 4.5.4) still did not find acceptance among the project participants, see section 5.3. As they did not see the comparison with the year before, the critique was the same: The *LearningLog* was discarded as being too long, not helpful and too tedious to fill in. The conclusion was that the **purpose** of documenting learning behaviour was not valued by the participants because the *LearningLog* in its previous form did not provide any feedback. The remedy was to design a new log that had the potential **characteristic** for immediate feedback, and did not weigh heavily on students' time. The downside was that these kinds of logs would put less emphasis on documenting learning behaviour.

The idea for a smartphone app was considered, but in the end a web page was settled upon. It was set up to collect the following data:

- participation code (shortened name and last digits of the matriculation number), to be saved for the next session

- logging period (free choice between current day or week)

- time spent learning for mathematics (slider, set on five hours for a day, variable between zero and ten hours a day, or five times this number for a week)

- percentage of planned workload accomplished (slider, set on 50%, variable between 0% and 100%)

- estimated present motivation for mathematics in percent (slider, set on 50%, variable between 0% and 100%)

- contentedness with learning success (4-point Likert scale from *not at all* to *very much*)

- effort exercised in learning mathematics (4-point Likert scale from *not at all* to *very much*)

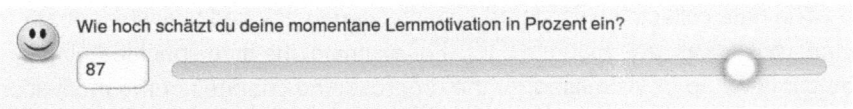

Figure 4.10: Excerpt from the *LearningLog online* from the Third Project Cycle, the Smiley Varies According to its Position.

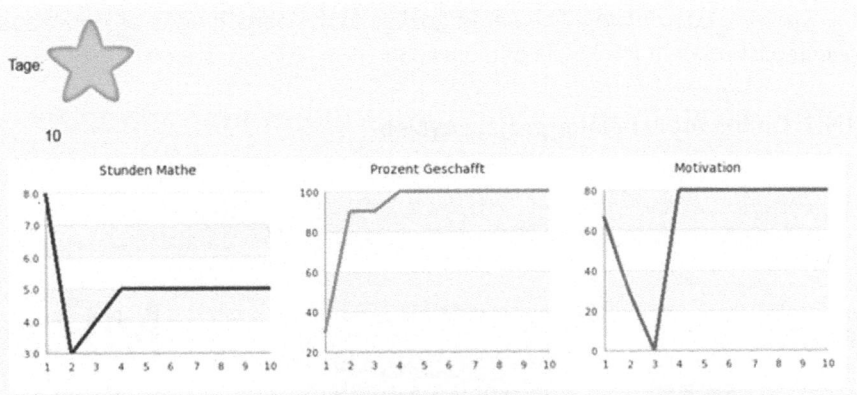

Figure 4.11: Feedback Provided by the *LearningLog online* From the Third Project Cycle.

This meant filling in was possible in a few minutes, for example when waiting for a train or on the bus. The webpage was not overfull and could easily be called up with a smartphone or tablet computer. In contrast to the *Learning Diary* or the *LearningLog* which had been rather bleak questionnaires to be completed, the *LearningLog online* was designed to please: The slider questions for time, percentage of workload completed and motivation were endorsed with symbols that varied according to the position of the slider, see Figure 4.10[30].

---

[30] Translation: How do you rate your motivation at the moment in percent?

The data collected was stored and intended to be used to give feedback in numerous forms, see Figure 4.11[31]. For example, the time spent learning can be summed up or visualised; or the progress and changes of the motivation level or the contentedness can be depicted as a graph. It is also possible to add more levels of feedback by comparing the time spent learning with a goal set at the beginning of the semester (or with a target value set by the lecturer, or with the average time calculated out of the times of all the students using the *LearningLog online*). The same applies to the percentages of workloads accomplished or to the levels of motivation.

4.6.2 Online tools in later project cycles

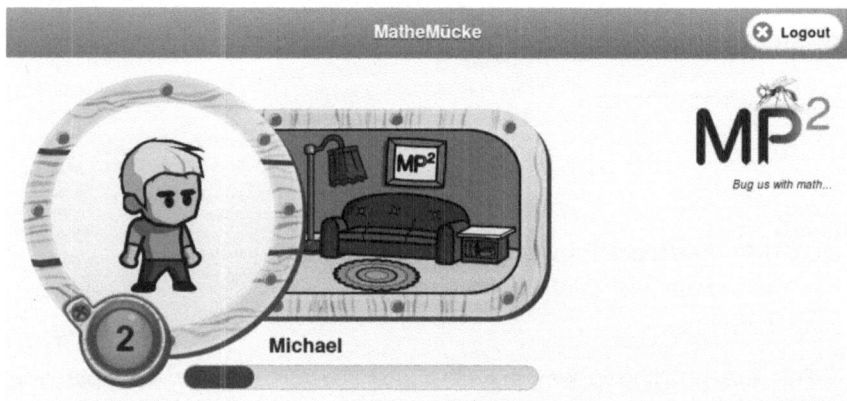

Figure 4.12: Screenshot from *MatheMücke*, Which Replaced the *LearningLog online* in the Fourth Project Cycle.

In order to elaborate on the gaming **characteristics** of an online tool, the mobile webside *MatheMücke*[32] was created for the fourth project cycle, see Kallweit and Griese (2014). The cover page contained an avatar, a background (both variable in different degrees as the gamer progressed), a level counter, a

---

[31] The yellow star shows that ten days (level 1) have been logged. The graphics visualise the hours spent on learning mathematics, the percentage of learning accomplished, and the changes in motivation.

[32] German *Mücke* means gnat, taken up in the motto "Bug us with math".

Figure 4.13: *MatheMücke* Avatars Depicting the Learner Type, from Unspecified to Visual, Auditive, Haptic-motoric, and Communicative (left to right).

progress bar, and links to different tasks or prizes / badges, see Figure 4.12 and Figure 4.13. The graphics were kept in modern comic style[33]. The possibilities for individual fit, structured learning assistance and useful tasks with feedback on individual mistakes were expanded.

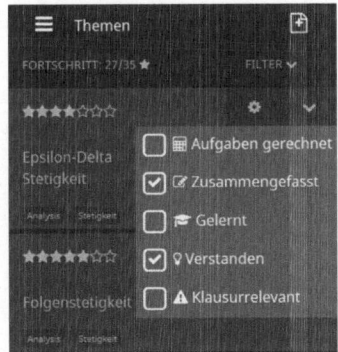

Figure 4.14: Screenshot from *MathePlus Companion*, Showing Flashcard Topics and their Attributes.

Another attempt to incorporate gaming elements was followed in the subsequent fifth project cycle: the web application *MathePlus Companion*, see Kiss and Kallweit (2015). The emphasis was shifted back to support in self-organisation, self-regulation, and reflection, with a concentration on mathematical support, e.g. via flashcards that can be ticked with *calculated tasks*,

---

[33] Created by Thorsten Kiss.

summarised, *memorised*, *understood*, and *relevant for the examination*, the last entailing a change in colour (see Figure 4.14).

4.6.3 Focus on facebook

At the time facebook was the most popular social network among the target group. It was therefore decided to abandon Google+ and Twitter and to realise the **procedure** as a closed group on facebook. The status of a group like this meant students had to ask for membership and be accepted by the group administrators, the *Learning Group* tutors. Though not every single student was on facebook, this was widely accepted (see Figure 4.15); the closed group had more than 60 members. For those who wanted to participate in the uploaded files but who did not want to use facebook, a dropbox[34] was established. This was actually the suggestion and the initiation of one of the participants, due to the **characteristic** that they were to organise their own learning.

Figure 4.15: Parting Words in the MP^2-Math/Plus Facebook Group from the Third Project Cycle.

---

[34] www.dropbox.com

5 Evidence Level: Methodology and Results

This chapter contains empirical explorations of the data collected in connection with MP^2-Math/Plus. It thus complements the previous work from Chapter 3 and Chapter 4, see orientation figure below.

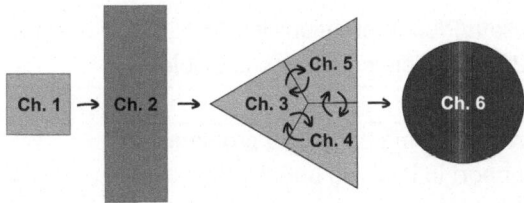

The empirical evidence starts with investigations of students' applications to MP^2-Math/Plus (section 5.1), a detailed description of the sample (section 5.2), and students' views on the project (section 5.3). A look at the attrition rates of *Mechanical Engineering* students (section 5.4) completes the first group of explorations. The examination statistics (section 5.5) cover both gender aspects and participation in the project. Before learning strategies are entered into linear models predicting academic success (section 5.7), the structure and viability of the data collected is explored (section 5.6). Then, the development of learning strategies (section 5.8) is scrutinised, with special emphasis on the influence of MP^2-Math/Plus. The evaluation of MP^2-Math/Plus on the basis of the empirical explorations presented here and of the theoretical considerations from Chapter 2 and Chapter 3 can be found in Chapter 6.

5.1 Applications for MP²-Math/Plus

Students had to fill in an online application form in order to take part in MP²-Math/Plus. The form required name, given name, matriculation number, result in the mini exam, year of school leaving examination, advanced or basic mathematics at school, advanced or basic (or no) physics at school, times available for the group meetings, and (most importantly) the reasons for wanting to participate in MP²-Math/Plus and a description of their problems in mathematics. Answers to the last two questions from the first project cycle were categorised as presented in Table 5.1 and Table 5.2, more than one category was allowed for each free text answer.

Table 5.1: MP²-Math/Plus Applications in 2010/2011, Reasons for Participation, $N = 209$, multiple categories possible

Category	#	%
Desire to overcome mathematical problems	171	82%
Need for support in learning behaviour	51	24%
Differences between school and university	14	7%
Motivation problems	10	5%
General help needed	7	3%
Not specified	4	

Table 5.2: MP²-Math/Plus Applications in 2010/2011, Description of Problems in Mathematics, $N = 209$, multiple categories possible

Category	#	%
Understanding and transfer	66	32%
Lack in basic skills	59	28%
Speed and routine	53	25%
Learning behaviour	52	25%
Specific knowledge gaps	48	23%
Not specified	9	

The statistics are not very different in the following project cycles; as an example the categorisation for the third project cycle is shown in Tables 5.3

and 5.4. The overwhelming desire to get mathematical support can again be identified. Students often express the hope to find contacts who will explain mathematical subject matter in a trustful atmosphere and in a way they will understand, e.g. "And I hope that, in case I have questions about specific topics, that someone can calmly explain everything to me" or "I hope that I get things explained that I didn't understand during the lecture, and that I can then reliably solve the assignments"[1].

Table 5.3: MP²-Math/Plus Applications in 2012/2013, Reasons for Participation, $N = 130$, multiple categories possible

Category	#	%
Desire to overcome mathematical problems	82	63%
Need for support in learning behaviour	31	24%
Differences between school and university	4	3%
Motivation problems	6	5%
General help needed	10	8%
Not specified	12	

Table 5.4: MP²-Math/Plus Applications in 2012/2013, Description of Problems in Mathematics, $N = 130$, multiple categories possible

Category	#	%
Understanding and transfer	29	22%
Lack in basic skills	10	8%
Speed and routine	13	10%
Learning behaviour	19	15%
Specific knowledge gaps	21	16%
Not specified	9	

The statistics for the other items, e.g. if MP²-Math/Plus applicants had an advanced or a basic mathematics course when at school, reveal that there is a

---

[1] Originally „Zudem hoffe ich, dass wenn ich Fragen zu bestimmten Themen habe mir jemand mit Ruhe alles erklären kann" or „Ich hoffe, dass ich Dinge erklärt bekomme die ich in der Vorlesung nicht verstanden habe und somit die Übungsblätter sicherer lösen kann", taken from the third project cycle.

Table 5.5: Gender Distribution

	Project Cycle	Male	Female	n
MP^2-Math/Plus	Second	62.07%	37.93%	58
Complete sample		84.57%	15.43%	188
MP^2-Math/Plus	Third	58.57%	41.43%	70
Complete sample		67.17%	32.83%	399
MP^2-Math/Plus	Fourth	55.97%	44.03%	134
Complete sample		78.96%	21.04%	908
MP^2-Math/Plus	2nd-4th	58.02% (152)	41.98%	262
Complete sample		76.52%	23.48%	1495

Additional data used for MP^2-Math/Plus participants,
not available for the first project cycle

higher proportion of students with only a basic course in mathematics among the applicants (60.47% in the third project cycle, 47.13% in the fourth cycle) when compared to the corresponding numbers from section 5.2 (46.80% respectively 33.67%, see Table 5.6[2]), although the proportion varies considerably from one cycle to the next.

These findings underpin the project **characteristic** to refer to the specific and current subject matter when introducing a learning method. They also stress the necessity of a competent helpdesk where students can have their questions answered and thus match the assessment from section 5.3 where students expressed their satisfaction with the MP^2-Math/Plus *HelpDesk*.

Nevertheless, the question remains if advertising and recruitment for MP^2-Math/Plus has appealed to the target group: first-year students who have realised their learning strategies are deficient and who are prepared to make an effort to remedy this problem. The low percentage of students explicitly referring to learning behaviour either as a reason for wanting to participate or as a central problem in their learning of mathematics casts some doubt on this issue. Numerous statements like "I would like to pass the examination in

---

[2] The percentages calculated from the applicants' data is more reliable than the data used for Table 5.6, as the latter only covered students present in the lecture hall during the survey.

Table 5.6: Students with Advanced or Basic Mathematics at School

	Project Cycle	Advanced Mathematics	Basic Mathematics	n
MP^2-Math/Plus	First	50%	50%	6
Complete sample		67.57%	32.43%	111
MP^2-Math/Plus	Second	27.27%	72.73%	11
Complete sample		66.67%	33.33%	162
MP^2-Math/Plus	Third	46.15%	53.85%	52
Complete sample		53.20%	46.80%	391
MP^2-Math/Plus	Fourth	60%	40%	95
Complete sample		66.33%	33.67%	897
MP^2-Math/Plus	1st-4th	53.05% (87)	46.95%	164
Complete sample		63.16%	36.84%	1561

maths 1" and "I hope to pass the maths exam"[3] hint at the fact that students either have not contemplated much about the kind of problems they have with university mathematics, or they are not accustomed to verbalising this kind of considerations. Both options support their accommodation into MP^2-Math/Plus, which will promote discussion on this matter.

In sum, the applicants for MP^2-Math/Plus express considerable problems with mathematical subject matter and a strong wish to overcome these.

5.2 Sample Description

RUB is a modern university with faculties covering humanities, social sciences, natural sciences, medicine, and engineering[4]. In the winter semester 2015/2016, more than 42,000 students were enrolled at RUB, 49% of them female. The majority of students (59%) have chosen humanities or social sciences, the next biggest groups study engineering (17%) and natural sciences (16%), medicine is rather small with 6%. The remaining 2% are involved in

---

[3] Originally „Ich würde gern die Mathe 1 Klausur bestehen" or „Ich verspreche mir die Mathe Klausur zu bestehen", taken from the third project cycle.
[4] rub.de/universitaet/fakultaeten

Table 5.7: Distribution of School Marks in Mathematics

	Project Cycle	Very good or good	Satisfactory or lower	n
MP^2-Math/Plus	First	0%	100%	6
complete sample		49.11%	50.89%	112
MP^2-Math/Plus	Second	27.27%	72.73%	11
complete sample		44.38%	55.63%	160
MP^2-Math/Plus	Third	44.23%	55.77%	52
complete sample		46.49%	53.51%	385
MP^2-Math/Plus	Fourth	51.61%	48.39%	93
complete sample		51.34%	48.66%	894
MP^2-Math/Plus	1st-4th	45.68% (74)	54.32%	162
complete sample		49.26%	50.74%	1551

other institutions[5]. In absolute numbers, more than 7,000 students are enrolled in one of the many engineering courses (there are six different Bachelor and eleven Master courses in engineering), of which RUB offers a wide choice with various major fields.

The engineering students in the focus of MP^2-Math/Plus are first-years studying Mechanical Engineering (*Maschinenbau*, MB), Civil Engineering (*Bauingenieurwesen*, BI) or Environmental Engineering and Resource Management (*Umwelttechnik und Ressourcenmanagement*, UTRM), which counted 713 when the project started, in the winter semester 2010/2011[6], and 706 five years later. Some of these may not actively follow an academic education, which would reduce the number of attendants in the mathematics lecture, but on the other hand, this number is augmented by those students who did not pass the examination in the year before.

The composition of the students participating in MP^2-Math/Plus is the focus of this section. This refers particularly to their educational background. Therefore the percentage of students with advanced mathematics at school,

---

[5] The numbers stem from http://www.rub.de/universitaet/fakten/menschlich/index.html, 08/05/2016.

[6] http://dwh.uv.ruhr-uni-bochum.de/aufgaben/planung-controlling-berichtswesen/statistik/archiv

Table 5.8: Type of School Visited

	Project Cycle	Gymna-sium	Gesamt-schule	other	n
MP²-Math/Plus	First	83.33%	0%	16.67%	6
complete sample		77.68%	15.18%	7.14%	112
MP²-Math/Plus	Second	81.82%	18.18%	0%	11
complete sample		87.73%	8.59%	3.68%	163
MP²-Math/Plus	Third	57.69%	36.54%	5.77%	52
complete sample		69.11%	23.29%	7.59%	395
MP²-Math/Plus	Fourth	67.71%	20.83%	11.46%	96
complete sample		72.67%	18.00%	9.34%	878
MP²-Math/Plus	1st-4th	66.06% (109)	24.85%	9.09%	165
complete sample		73.71%	18.15%	8.14%	1548

the distribution of school marks, the type of school[7] visited, the rating of their school education as appropriate preparation for university, and attendance of the preparatory mathematics course at university will be investigated. Furthermore the distribution of gender, of age, and of the first language are of interest.

All numbers are given for MP²-Math/Plus participants and for the complete sample of first year engineering students (as far as they participated in the survey). Whenever students are described as participants in MP²-Math/Plus, this refers to the SLG (see section 4.3.1), as in later project cycles, this was the only project group. The paper-and-pencil survey was conducted at the beginning of the first semester, in the mathematics lecture. The variation of n stems from missing data.

Due to the often small numbers, variations of the percentages are mostly not significant. In some cases, however, the observed numbers lie outside the 2σ-confidence interval, which covers 95.4% of the possible values, or even outside the 3σ-confidence interval, which stands for 99.7%. The greatest

---

[7] Germany, or more specifically North Rhine-Westphalia, offers different schools in which to attain a school leaving certificate that allows to attend university: the traditional Gymnasium is the most popular, with the comprehensive Gesamtschule gaining ground in recent years. There are also other (rare) possibilities to qualify for university education.

Table 5.9: School Education Rated as Appropriate for University

	Project Cycle	1	2	3	4	5	n
Math/Plus	First	0%	16.67%	66.67%	16.67%	0	6
complete		10.71%	20.54%	34.82%	25%	8.93%	112
Math/Plus	Second	9.09%	0%	72.73%	18.18%	0%	11
complete		8.86%	28.48%	53.16%	9.49%	0%	158
Math/Plus	Third	1.89%	24.53%	56.60%	16.98%	0%	53
complete		7.25%	41.25%	37.25%	14.25%	0%	400
Math/Plus	Fourth	5.10%	40.82%	35.71%	18.37%	0%	98
complete		8.01%	36.36%	37.71%	18.02%	0%	899
Math/Plus	1st-4th	4.17%	32.14%	45.83%	17.86%	0%	168
complete		8.09%	35.63%	38.94%	16.70%	0.64%	1569

From 1 = very good preparation to 5 = very bad preparation

difference can be found in the distribution of gender, see Table 5.5. For $p = 0.77$ (because 77% of the students are male) and $n = 262$, $\sigma \approx 6.81 > 3$, and the 3σ-confidence interval is $[182; 222]$; obviously the observed number of males participating in MP^2-Math/Plus, 152, lies far outside. Females take part in MP^2-Math/Plus in a much higher proportion than expected, considering their percentage of the students as a whole.

Also, on the whole, MP^2-Math/Plus participants are less likely to have attended an advanced course in mathematics at school, see Table 5.6 ($p = 0.63$ and $n = 164$ satisfy $3 < \sigma \approx 6.18$, and the 2σ-interval is $[91; 115] \not\ni 87$). The school marks in mathematics do not vary significantly, however, see Table 5.7. The type of school visited shows differences: MP^2-Math/Plus participants are slightly less likely to have attended the traditional *Gymnasium*, see Table 5.8 ($p = 0.74$, $n = 165$, $\sigma \approx 5.63 > 3$; 2σ-interval $[111; 133] \not\ni 109$).

The survey also contained a Likert-scale question if the students felt school had prepared them adequately for university. The hypothesis was that project participants would feel significantly less well-prepared, which might be one reason they had applied. But this could not be verified by the collected data, see Table 5.9; participants felt as well- or badly-prepared as the complete sample.

Table 5.10: Attendance of Preparatory Mathematics Course

	Project Cycle	Attended Prep. Course	Did not attend	n
MP^2-Math/Plus	First	100%	0%	6
complete sample		72.32%	27.68%	112
MP^2-Math/Plus	Second	72.73%	27.27%	11
complete sample		37.65%	62.35%	162
MP^2-Math/Plus	Third	52%	48%	50
complete sample		43.00%	57.00%	393
MP^2-Math/Plus	Fourth	62.89%	37.11%	97
complete sample		54.99%	45.01%	902
MP^2-Math/Plus	1st-4th	61.59% (101)	38.41%	164
complete sample		51.43%	48.57%	1569

There was an indication that MP^2-Math/Plus participants wanted more help than the complete group: They attended the preparatory course offered by the university before lectures started in a much higher proportion, see Table 5.10 ($p = 0.51$, $n = 164$, $\sigma \approx 6.40 > 3$, 2σ-interval $[71; 96] \not\ni 101$). This suggests that project participants did feel unsure about mathematics even before they had gathered their first experiences with university mathematics.

The first language spoken by the students participating in MP^2-Math/Plus was more likely not to be German than the average would suggest, see Table 5.11 ($p = 0.76$, $n = 163$, $\sigma \approx 5.45 > 3$, 2σ-interval $[113; 134] \not\ni 112$).

Exploration of the distribution of age showed no differences, for the complete sample $M \approx 20$ and $\sigma \approx 2.39$ (in years) were calculated. The variations were only to be found in the decimals; the median varied between 20 and 21, depending on the group investigated.

It can be summarised that the students starting an engineering course at RUB are homogeneously young and feel prepared for university in an un-remarkable way. MP^2-Math/Plus project participants have a slightly weaker educational background, although this cannot be manifested in school marks, but in lower attendance of a *Gymnasium* and of an advanced course in mathematics at school. Project participants are disproportionately often female and have visited a preparatory course more often. They are slightly more likely to have a non-German background.

Table 5.11: First Language

	Project Cycle	German	Other	n
MP^2-Math/Plus	First	100%	0%	6
complete sample		89.19%	10.81%	111
MP^2-Math/Plus	Second	81.82%	18.18%	11
complete sample		88.27%	11.73%	162
MP^2-Math/Plus	Third	70.59%	29.41%	51
complete sample		77.35%	22.65%	393
MP^2-Math/Plus	Fourth	64.21%	35.79%	95
complete sample		71.69%	28.31%	890
MP^2-Math/Plus	1st-4th	68.71% (112)	31.29%	163
complete sample		76.09%	23.91%	1556

5.3 Students' Views on MP^2-Math/Plus

At the end of the semester, when the MP^2-Math/Plus participants had experienced all project procedures, they were asked to fill in an evaluation questionnaire to express their opinion on the project and its procedures. The questionnaire used in the first project cycle in 2010/2011 can be found in Appendix C, the surveys for the subsequent cycles were compiled in a similar fashion, i.e. the numbers and sometimes the names of the procedures changed, but the pattern of questions (rating questions, followed by *What could be improved?*, *Would you recommend xy?*, and *Would you like to tell us something else about xy?*) stayed the same. Some answers seemed cautious, as the examination had not yet taken place, and there was still considerable insecurity about the level of its challenge, e.g. "It remains to be seen how my examination turns out"[8]. From the second project cycle on, the assessment surveys were not only anonymous, but also lacked a participation code.

Figure 4.2 presents the answers to the question if the participants would recommend the various interventions and was discussed in detail in section 4.4. In sum, participants approved of MP^2-Math/Plus but disapproved of the daily learning diary (see section 4.3.4), which was changed into a weekly *LearningLog* (see section 4.5.4) in the second project cycle. The second cycle

---

[8] Originally: „Es wird sich zeigen, wie meine Matheklausur ausfällt."

contained more procedures which were more closely linked, see section 4.5.2. The results of the analogous questions for the second project cycle are given in Figure 5.1. Again, MP²-Math/Plus has met with general approval among its participants. As in the first project cycle, the university helpdesk is assessed as less recommendable than the MP²-Math/Plus *HelpDesk*. The newly introduced procedures *LearningLog* and *Workbooks* met with mixed acceptance, particularly the rating for the *LearningLog* prompted further changes concerning this procedure, see section 4.6.2. The MP²-Math/Plus *HelpDesk* and the redesigned *Learning Groups* scored best. The comments in the open questions corroborate this overall positive impression, e.g. "A truly great project which helps to come to grips with initially irresolvable tasks and problems and later even understand them", "It really helped me very much, first of all the obligation to attend all meetings is helpful for your weaker self, so that you really go there regularly, and the revision course is worth a mint", or "[It] was very helpful, and you learn a lot for other subjects, too"[9]. Of course, critique was also voiced, e.g. "The project started very late", "drop the LearningLog", and "more appointed times for the [MP²-Math/Plus] helpdesk"[10].

The analogous numbers from the fourth[11] project cycle are given in Figure 5.2. MP²-Math/Plus again enjoyed approval by its participants. Those interventions that were not used by a considerable number of students[12] are understandably rated lower. Here, the differences between the university's SZMA helpdesk and the MP²-Math/Plus *HelpDesk* have disappeared. This is probably due to an initiative to improve the university helpdesk by offering workshops for the helpdesk staff, especially assembled learning summaries on all relevant topics, an online collection of anonymised old examination papers, and concise revision[13], see Buchsteiner and Kallweit (2015).

---

[9] Originally „Eine wirklich tolle Einrichtung, die einem hilft, mit anfangs unlösbaren Aufgaben und Problemen klar zu kommen und dann sogar zu verstehen", „Es hat mir wirklich sehr geholfen, vorallem die Pflicht zu den Veranstaltungen ist hilfreich für seinen inneren Schweinehund, dass man dort auch regelmäßig hingeht und das Repititorium ist Goldwert (sic)", and „War sehr hilfreich und man lernt viel auch für andere Fächer", taken from the second project cycle.

[10] Originally „Das Projekt hat sehr spät angefangen", „LearningLog streichen", and „mehr HelpDesk-Termine", taken from the second project cycle.

[11] There is no data from the third project cycle.

[12] The dropbox had originated in one of the learning groups and was not propagated much among the others. The same applied to the facebook group which furthermore may have suffered from misgivings against facebook that came up in the 2010s.

[13] *30-Minute-Maths*, a regular half-hour session concentrating on one topic.

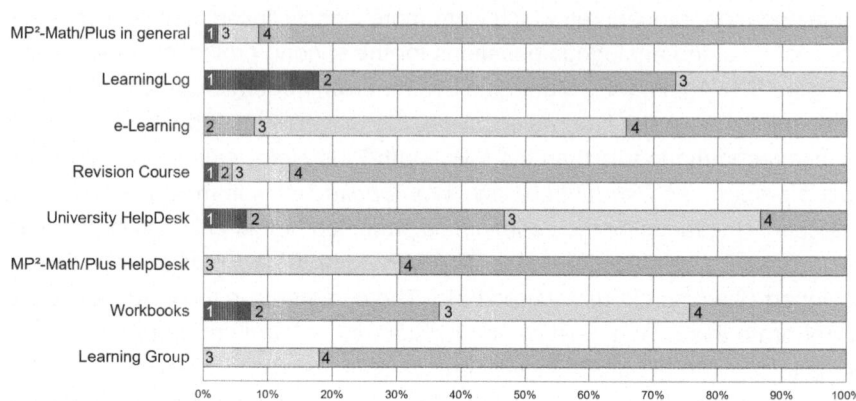

Figure 5.1: Second Project Cycle, *Would you recommend the intervention?*
Likert scale from 1 (not at all, red) to 4 (very much, green), $n = 48$.

Table 5.12: Rating of Methodical Support

Project Cycle	Very bad (1)	Bad (2)	Good (3)	Very good (4)	Mean	SD
First	0	2	17	15	3.38	0.59
Second	0	0	15	25	3.63	0.48
Fourth	2	18	41	42	3.19	0.79

The conceptual **characteristic** of MP^2-Math/Plus is the concentration on learning strategies. It is therefore of interest to explore how the participating students rated the support in this area. This was covered by the question *How do you rate the methodical support?*, see Table 5.12. To better assess these numbers, the answers to *How do you rate the subject-specific support?* are given in Table 5.13 for comparison.

It is difficult to draw conclusions from these numbers, as the differences are small and not significant, although the ratings for subject-specific support appear uniformly better. These two questions may also reflect participants' needs for methodical or subject-specific support, maybe more so than the actual support given.

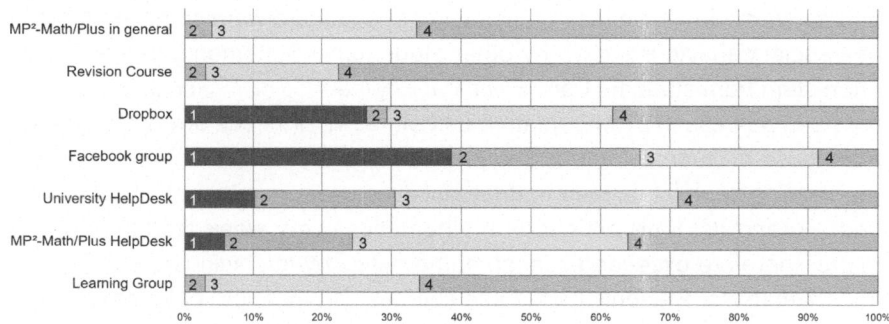

Figure 5.2: Fourth Project Cycle, *Would you recommend the intervention?*
Likert scale from 1 (not at all, red) to 4 (very much, green), *n* = 112.

Table 5.13: Rating of Subject-specific Support

Project Cycle	Very bad (1)	Bad (2)	Good (3)	Very good (4)	Mean	SD
First	0	0	8	25	3.76	0.43
Second	0	0	11	29	3.73	0.45
Fourth	0	13	37	56	3.41	0.70

A last table reveals the answers to the question *Did participating in MP²-Math/Plus help you?*, see Table 5.14. On the whole, 95% of MP²-Math/Plus participants (170 of 179) state that they found the project helpful or even very helpful. Some of the verbatim statements back this impression, e.g. "Great project. Helps first-years to get started with the challenging mathematical material. And you realise that you are not all alone. There are many others who don't understand the subject-matter either, and that helped me tremendously to be more confident concerning the tasks and the exam"[14]. Particularly in the fourth project cycle, though, there was a considerable number of comments demanding to be actively spoon-fed subject-specific assistance (while they themselves would remain rather passive), like demonstrations of exercises

---

[14] Originally „Tolles Projekt. Hilft den neuen beim Einstieg in den doch sehr anspruchsvollen Mathestoff. Ausserdem erkennt man dort das (sic) man nicht ganz alleine ist. Es gibt auch viele andere die den Stoff nicht verstehen und das half mir ungemein ein bisschen selbstbewusster zu sein was die Aufgaben und Klausur anging", taken from the fourth project cycle.

and ready-made summaries or instructions, and there were complaints about anti-social behaviour shown by other students[15]. Statements of this nature often stem from students from *Electrical Engineering* or *IT Security*, who had not been addressed by MP^2-Math/Plus before. Their needs obviously differed from those of students of *Machine Engineering*, *Civil Engineering*, or *Environmental Engineering*, as their mathematics lecture was more advanced: They had six (and not four) periods of mathematics every week, and the subject matter therefore exceeded school mathematics by far, including e.g. Fourier transformations. Students from these courses had therefore been assigned to their own *Learning Groups*. Their criticism implied project re-design specifically for this group of students, for example with a stronger emphasis on subject matter. This was attempted with the MP^2-Math/Plus *Companion*, see section 4.6.2.

Table 5.14: Rating of Helpfulness of MP^2-Math/Plus

Project Cycle	very bad (1)	bad (2)	good (3)	very good (4)	Mean	SD
First	0	0	12	20	3.53	0.61
Second	1	1	10	36	3.69	0.62
Fourth	0	7	40	52	3.45	0.62

In sum, participating students rated MP^2-Math/Plus favourably, with the exception of the learning diary and its successors. The appreciation of the usefulness of learning strategies is inconclusive.

5.4 Attrition Rates

This section explores how students participating in MP^2-Math/Plus fare in the course of their further studies. The idea behind this is, that someone who

---

[15] "We didn't do enough calculations, the time in the learning groups should have been used more efficiently" or "Participants' social behaviour should be required as almost all participants were anxious for their personal benefit", originally „Es wurde zu wenig aktiv gerechnet, die Zeit in den Lerngruppen hätte man dementsprechend effizenter nutzen können"and „man sollte das sozialverhalten der teilnehmer stärker fordern da annähernd alle Teilnehmer nur auf ihren persönlichen Vorteil bedacht waren (sic)", by an *Electrical Engineering* and an *IT Security* student from the fourth project cycle.

has learned about learning strategies in reference to mathematics may be successful in other subjects, too.

Table 5.15: Retention Statistics for *Mechanical Engineering* Students from the First and Second Project Cycle

Group	Complete Data Set	SLG only	SLG and SDG
n	2056	33	71
still enrolled	1531	19	45
percentage	78%	58%	63%

In our case, it was possible to retrieve data for students of *Mechanical Engineering* who started their university course in the winter semester of 2010 / 2011 or of 2011 / 2012, when the first and the second project cycle of MP^2-Math/Plus were under way. Enough time has passed for this cohort to have attempted (and re-attempted) all three examinations in mathematics and the two examinations in mechanics, which are considered as similarly challenging. The data comprises 2056 students, 33 of which have taken part in MP^2-Math/Plus in the *Supported Learning Group* (SLG)[16], and 38 in the *Self-Directed Group* (SDG) which was well-established in the first project cycle. Table 5.15 shows how many of these were still enrolled at RUB in 2015.

The same statistics are presented in Table 5.16 for students participating in MP^2-Math/Plus. SLG has an overall attrition rate of 25%, SLG and SDG together yield 24%. In essence, the numbers do not vary noticeably between MP^2-Math/Plus participants and the complete group, as Figure 5.3 visualises, where the two circle diagrams look identical although they are based on the complete cohort (left) and on the MP^2-Math/Plus participants (right).

Next, the numbers of students passing their mathematics and mechanics examinations are scrutinised. Ideally, a student should pass the examinations in *Mathematics 1* and *Mechanics A* in the first semester, *Mathematics 2* and *Mechanics B* in the second semester, and *Mathematics 3* in the third. The data at our disposal makes allowances for examinations that are taken later, too, because for a degree, it does not matter when an examination was taken.

---

[16] The seemingly small numbers are due to the fact that only data for students of *Mechanical Engineering* were available.

Table 5.16: Attrition Statistics for *Mechanical Engineering* Students from
the First and Second Project Cycle, SLG only (SLG and SDG)

Course Module	# Pass	Attrition Rate to Next Module
Mathematics 1	20 (49)	40% (45%)
Mechanics A	8 (22)	88% (77%)
Mathematics 2	7 (17)	71% (82%)
Mechanics B	5 (14)	100% (86%)
Mathematics 3	5 (12)	

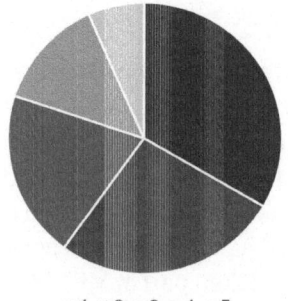 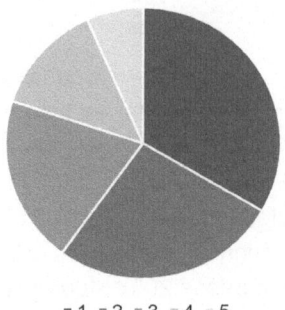

■1 ■2 ■3 ■4 ■5   ■1 ■2 ■3 ■4 ■5

Figure 5.3: *Mechanical Engineering* Students' Attrition,
shares having passed only one module in *Mathematics 1* (1),
two modules, *Mathematics 1* and *Mechanics A* (2),
three modules, *Mathematics 1* and *2* and *Mechanics A* (3),
four modules, *Mathematics 1* and *2* and *Mechanics A* and *B* (4),
or all five modules (5);
complete data (left), SLG and SDG (right)

It is thinkable to either look at the attrition rates when students pass from one semester to the next, or first through one subject and then through the other. The numbers for the first approach are given in Table 5.17, as they reflect the progress through the course authentically. The overall attrition rate of the students covered by the data set is 31%, i.e. this is the percentage of *Mechanical Engineering* first-year students from 2010 / 2011 or 2011 / 2012 who have passed all three mathematics and the two mechanics modules.

Table 5.17: Attrition Statistics for *Mechanical Engineering* Students from the First and Second Project Cycle

Course Module	# Pass	Attrition Rate to Next Module
Mathematics 1	996	48%
Mechanics A	474	87%
Mathematics 2	411	76%
Mechanics B	314	98%
Mathematics 3	308	

5.5 Examination Statistics

The professed goal of MP^2-Math/Plus is to support students in their first semester at university, by avoiding unnecessary dropout – which is often connected with examination failure, see Heublein and Barthelmes (2010) or Heublein et al. (2012). Therefore, the pass rates and the marks in the written examination are *the* hard fact on which to measure the success of the project.

The data explored in section 5.2 has shown that MP^2-Math/Plus participants vary in some respects from the complete sample; e.g. they have a slightly weaker educational background. Therefore comparisons between project participants and non-participants must be viewed with this bias in mind. In some instances, participants will therefore be compared to a group of students with similar preconditions.

The findings concerning the **procedure** are **summarised** and referred to as **empirical arguments**, according to the *Design Research* structure by van den Akker (2013); this is indicated by bold face print. Section 5.5.3 contains an overall summary of the findings concerning examination statistics.

5.5.1 Gender aspects

In order to present the data acquired as clearly as possible, we opted for Venn diagrams for the three features examination success, project participation, and gender, as extension of the well-known contingency tables. The numbers are given for each subset of the Venn diagram (on the basis of the number of students attending the written examination, not on the basis of those who were enrolled in the course and would have been allowed to attend), the ones in brackets are the expected numbers, calculated on the basis of the overall percentages. All Venn diagrams are organised in the way that the top centre set contains the students who passed the exam, the bottom left set those who participated in MP^2-Math/Plus, and the bottom right set are the females. This means that outside the sets, the numbers for male non-participants who did not pass the examination is given, see Figure 5.4.

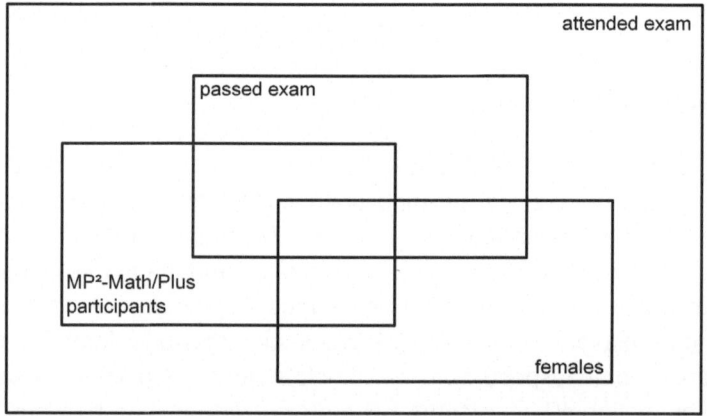

Figure 5.4: Structure of Venn Diagrams for Examination Statistics

For the third project cycle in 2012/2013, the Venn diagram is based on the numbers given in Table 5.18. The Venn diagram (see Figure 5.5), however, contains more information than the table. The expected numbers given in

Table 5.18: Examination Statistics for the Third Project Cycle in 2012/2013

	Pass	Fail	Sum
MP^2 Participants	51 (=28+23)	19 (=13+6)	70 (=41+29)
Non-Participants	331 (=254+77)	249 (=185+64)	580 (=439+141)
Sum	382 (=282+100)	268 (=198+70)	650 (=480+170)

The numbers in brackets are for males and females only, respectively.

brackets there allow for comparison between expected and genuine values, thus indicating that wherever there is a notable difference, there is need for further calculation and a reason for interpretation. If this is the case should be checked for all subsets, i.e. also for those consisting of two or more compartments of the diagram. At first glance, in Figure 5.5, regarding project participation and pass rate (28 + 23 = 51, 30 + 11 = 41 expected), there are notably more successful MP^2-Math/Plus participants than expected on the whole, too, but this is due to the females, as from male MP^2-Math/Plus participants, 30 were expected to pass, but only 28 did. The most notable differences (highlighted in bold) are for the central subset (23, 11 expected), containing female MP^2-Math/Plus participants who passed the examination, and for the subset on the left (13, 22 expected) containing male MP^2-Math/Plus participants who did not pass the examination. On the other hand, there were 100 (= 23 + 77) females who passed the examination, which is exactly the expected number (11 + 89).

This phenomenon can be **summarised** as such: When referring to examination success, female students seem to profit disproportionally more from MP^2-Math/Plus than males, although there is no general tendency for females to pass more often than males. This stands in contrast to the lower mathematics competence of girls as reported in section 2.1.1. In the third project cycle in 2012/2013, this picture was very clear, this **empirical argument** led to intensified efforts to meet the needs of male MP^2-Math/Plus participants, e.g. by choosing tutors and *Mentors* (see section 4.5.3) with regard to diversity of role models and their ability to relate to all kinds of personalities.

Before the third project cycle in 2012/2013, gender had not been a focus of the re-designed **procedure**, as the following numbers show. For the second project cycle in 2011/2012, this tendency is detectable (but rather weak) when comparing MP^2-Math/Plus participants to a group with comparable results in the first mini exam of the semester. The numbers are given in

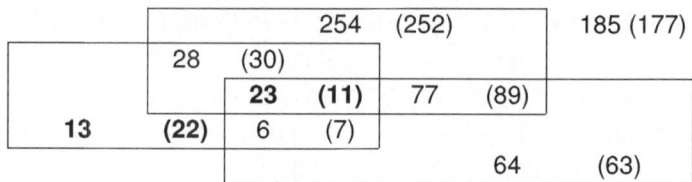

Figure 5.5: Venn Diagram for Examination Statistics for the Third Project Cycle in 2012/2013 (top centre set: passed exam, bottom left set: MP² participant, bottom right set: female)

Table 5.19: Examination Statistics for the Second Project Cycle in 2011/2012, MP² Participants Opposed to Comparison Group

	Pass	Fail	Sum
MP² Participants	41 (=25+16)	17 (=11+6)	58 (=36+22)
Comp. Group	310 (=262+48)	272 (=219+53)	582 (=481+101)
Sum	351 (=287+64)	289 (=230+59)	640 (=517+123)

The numbers in brackets are for males and females only, respectively.

Table 5.19 and in Figure 5.6. The highlighted numbers indicate that much less female MP²-Math/Plus participants failed the examination than expected (6, as opposed to 11 expected), although 22 (= 16 + 6) females participated in MP²-Math/Plus, exactly as expected (= 11 + 11). The impression that females profit more from MP² project interventions than males is even less pronounced when we oppose participants to non-participants, see Table 5.20 and Figure 5.7.

Table 5.20: Examination Statistics for the Second Project Cycle in 2011/2012, MP² Participants Opposed to Non-Participants

	Pass	Fail	Sum
MP² Participants	41 (=25+16)	17 (=11+6)	58 (=36+22)
Non-Participants	420 (=357+63)	302 (=242+60)	722 (=599+123)
Sum	461 (=382+79)	319 (=253+66)	780 (=635+145)

The numbers in brackets are for males and females only, respectively.

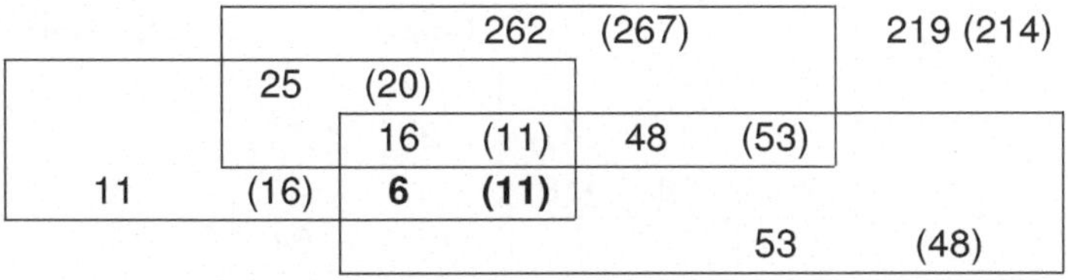

Figure 5.6: Venn Diagram for Examination Statistics for the Second Project Cycle in 2011/2012, MP^2 Participants Opposed to Comparison Group (top centre set: passed exam, bottom left set: MP^2 participant, bottom right set: female)

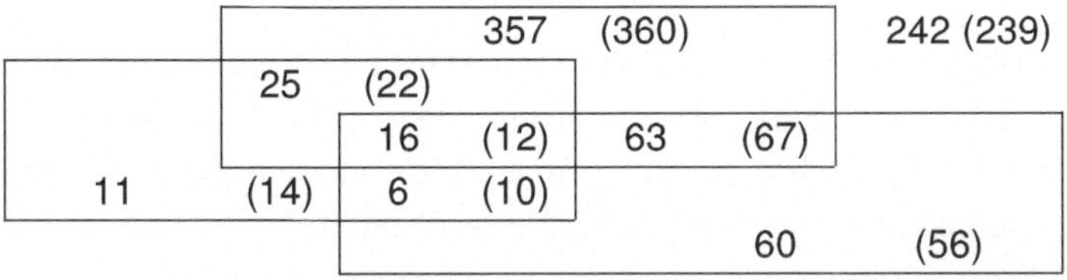

Figure 5.7: Venn Diagram for Examination Statistics for the Second Project Cycle in 2011/2012, MP^2 Participants Opposed to Non-Participants (top centre set: passed exam, bottom left set: MP^2 participant, bottom right set: female)

Table 5.21: Examination Statistics for the Fourth Project Cycle in 2013/2014, MP^2 Participants Opposed to Non-Participants

	Pass	Fail	Sum
MP^2 Participants	95 (=54+41)	39 (=21+18)	134 (=75+59)
Non-Participants	584 (=457+127)	308 (=254+54)	892 (=711+181)
Sum	679 (=511+168)	347 (=275+72)	1026 (=786+240)

The numbers in brackets are for males and females only, respectively.

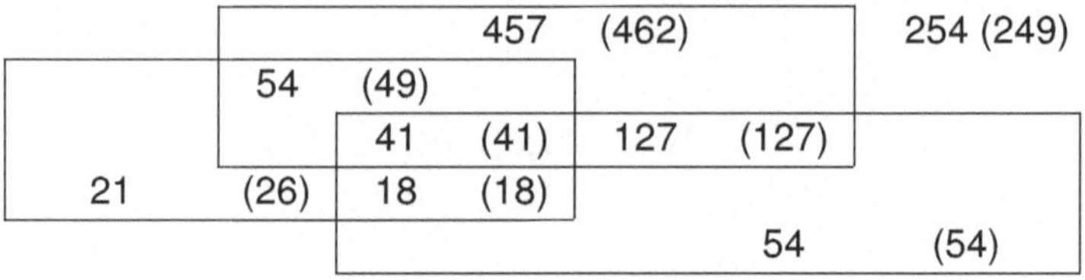

```
                        457    (462)        254 (249)
            54    (49)
                  41    (41)   127    (127)
      21    (26)  18    (18)
                               54         (54)
```

Figure 5.8: Venn Diagram for Examination Statistics for the Fourth Project Cycle in 2013/2014, MP^2 Participants Opposed to Non-Participants (top centre set: passed exam, bottom left set: MP^2 participant, bottom right set: female)

The question is, therefore, if the increased efforts to appeal to the males among the MP^2-Math/Plus participants were rewarded. To answer this question, Table 5.21 and Figure 5.8 give the numbers for engineering students from the different courses, and there are no remarkable differences at all. But this might be due to an tendency in one of the engineering courses being nullified by an opposite tendency in another course. So, it is necessary to explore the analogous statistics separately for the engineering courses concerned, see Tables H.1, H.2, and H.3, respectively Figures H.1, H.2, and H.3 in Appendix H.

Apart from the fact that these numbers carry the information that females enrol in environmental engineering rather than in IT security courses, **in sum** there is nothing left of the **empirical** tendency described above, namely that female participants in MP^2-Math/Plus profit more from the project interventions than males. Consequently, the chosen direction referring to the project **procedure** on e-learning was continued, by further developing the electronic tool, see section 4.6.2.

5.5.2 MP^2-Math/Plus project participation

The calculations referring to gender must not divert us from another important question, namely if there is **empirical** evidence that participation in MP^2-Math/Plus helped engineering students pass their written examination at all. The answer can be gathered from the numbers in the previous tables and figures in this chapter, for a closer view they are summarised in Table 5.22[17]. Now the impression described above, that the fourth project cycle in 2013/2014 can be regarded as a success in terms of males performing as well as females, becomes tinted, though: It appears that, on the contrary, females were rather performing as badly as the males, thus tarnishing the overall success rate. Considering that MP^2-Math/Plus participants show more problematic features in terms of previous education, the numbers can be rated in favour of the project concept, though.

Table 5.22: Summary of Examination Statistics, Pass Rates (SLG only)

Project Cycle, Year	MP^2-Math/Plus	Non-Participants
First, 2010/2011	48.78%	60.45%
Second, 2011/2012	70.69%	58.17%
Third, 2012/2013	72.86%	57.09 %
Fourth, 2013/2014	70.90%	65.47%

There is another indication of project success that can shed some light. When evaluating the data, it was a surprise to discover how many students did not attend the written examination at all. This even applied to students who participated in the project, meaning that they had taken part in the weekly sessions, had worked on the weekly assignments and generally had made an effort to master mathematics. The question remained why they would not go to the examinations, particularly when this means an unsuccessful attempt in their study record (unless they could provide an incapacity certificate). It is not possible to provide reliable numbers for an attendance rate for the examination for non-participants of MP^2-Math/Plus, as the reference quantity is unclear: it might be the number of students enrolled in the course, or the number of those who had actually started the course properly – but then the definition of *properly* poses problems: Have you started a university course properly if you hand in the first two weekly assignments, and nothing afterwards? It is possible, however, to calculate which proportion of MP^2-Math/Plus participants attended

---

[17] Although these are the facts that essentially determine the success of MP^2-Math/Plus, it has to be said that they do not represent objective measurement of students' academic performance. There is a consensus about what exactly should be dealt with in the mathematics lectures for engineering students, but the difficulty and nature of the tasks employed to attest these skills vary considerably: In some years, there are multiple choice tasks that bear their very own specifities (and at other times it is calculation tasks only), and the proposed pass barrier also varies slightly.

the written examination, especially as their *proper* participation in the project undergoes checking shortly before the written examination, when admittance to the revision course is determined.

The numbers are **summarised** in Table 5.23 and present an **empirical argument** that the **purpose** of ensuing commitment was accomplished, as attendance of the written examination among MP^2-Math/Plus participants is consistently high.

Table 5.23: Examination Attendance of MP^2-Math/Plus Participants

First	Second	Third	Fourth Project Cycle
<60%	>80%	81.40%	92.75%

Table 5.24: Absolute Numbers of MP^2-Math/Plus Participants with Examination Attendance, Only Machine, Civil and Environmental Engineering

First	Second	Third	Fourth Project Cycle
41	58	70	102

Table 5.25: Absolute Numbers of *Pass* MP^2-Math/Plus Participants, Only Machine, Civil and Environmental Engineering

First	Second	Third	Fourth Project Cycle
20	41	51	78

A third indication of project success is given by the absolute numbers of MP^2-Math/Plus project participants (respectively of those who passed the examination), see Tables 5.24 and 5.25. The advantage over pass rates here is that absolute numbers are not dependent on examination attendance. As from the fourth project cycle in 2013/2014 on, MP^2-Math/Plus was extended and more engineering courses were included, for a fair comparison of numbers, only those numbers are included that refer to the engineering courses that had been part of MP^2-Math/Plus from the beginning (*Machine Engineering*, *Civil Engineering*, and *Environmental Engineering*). Both tables show a clear upwards

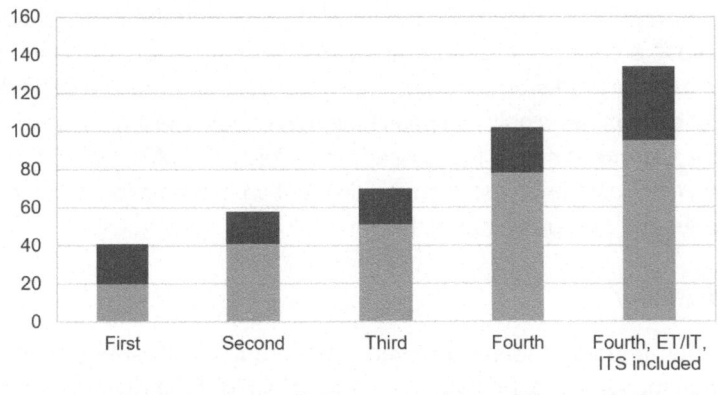

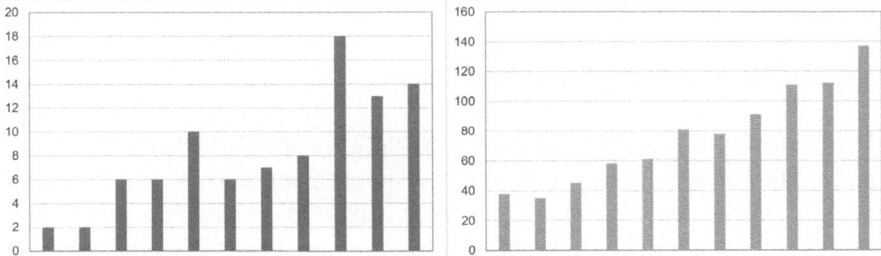

Figure 5.9: MP^2-Math/Plus Participants with Examination Attendance, by Project Cycle

Figure 5.10: Distribution of Marks among MP^2-Math/Plus Participants (left) and the Complete Sample (right), data from second and third project cycle, $n = 1439$

tendency, as **summarised** in Figure 5.9, entailing an **empirical argument** for the fact that MP^2-Math/Plus' acceptance was growing among students, and that the number of students who were able to benefit from the project's support was increasing every year.

Apart from the statistics for passing or failing the examination, it is also of interest how the marks are distributed among MP^2-Math/Plus participants and non-participants. This might show if the slightly weaker educational background is still empirically observable in the examination results. The distribution of marks is presented in Figure 5.10[18]. It is obvious that the examinations are not graded to a curve, but rather that the distribution has an increase towards the *pass* and *fail* marks on the right. This implies that there is a considerable number of students who only just pass their mathematics examination and as an **empirical argument** supports the need for a project like MP^2-Math/Plus. There is no significant difference between the distribution of marks for MP^2-Math/Plus participants and non-participants, however.

5.5.3 Summary

Temporarily, female students profited more from of MP^2-Math/Plus participation than males, although they did not show a higher ability in general. This tendency could not be observed in later project cycles, after the adaptation of some project procedures. On the whole, MP^2-Math/Plus participants show a high readiness to attend the written examination with satisfactory pass rates and average marks. Altogether, the acceptance of MP^2-Math/Plus has grown among engineering students.

5.6 Structure and Viability of LIST Data

The LIST data from four project cycles (2010/2011 until 2013/2014) offers many possibilities of exploration. To ensure a systematic approach as well as a both general and in-depth views, the plan is to explore the manifold data with the perspective of later answering research questions three to five (see section 3.3) in Chapter 6. The findings from this chapter will then be used to evaluate the development and influence of learning strategies as the **characteristic** of MP^2-Math/Plus as a *Design Research* project, according to van den Akker (2013).

The focus of this chapter is to explore the factor structure of LIST in order to test its worth for further analysis. To address this issue, there are several calculations to be done. First, Cronbach's α will yield information on the internal reliability of the proposed twelve scales in LIST, see section 5.6.1. Second, an exploratory factor analysis (EFA) will then reveal the loadings of the items on the scales, see section 5.6.2. Third, a confirmatory factor analysis (CFA) will clarify if the data actually fits the proposed factor model, see section 5.6.3. Although

---

[18] The marks range from 0.7 (excellent) to 4.0 (pass). The last column on the right represents *fail*.

the LIST questionnaire has proven its quality many times over, there is still the need to detect if it suits the **purpose** of MP²-Math/Plus as a *Design Research* project with its substantial emphasis on learning strategies. An additional fourth calculation will be done using the data from the individual surveys separately: The Cronbach's α values for each pre or post survey can bring to light in how far the scales are reliable to use for analysis or comparisons for single cohorts.

   As a considerable share of the calculations is of a rather technical nature, the findings will be summarised in section 5.6.4.

5.6.1 Internal reliability

The computations of Cronbach's α as a measure for internal reliability will be conducted using the data collected in three project cycles, from 2011/12 to 2013/2014. In each project cycle, LIST was used in a pre / post design: the same items were used at the beginning of the semester, and at the end (changed to the past tense), see section 2.2.2 and Appendices A and B. The first cycle, 2010/2011, is treated as a pilot year and therefore neglected here for more than one reason: There were problems matching the pre and post data, the Likert scales used five points instead of four in subsequent years, and the project procedures had not yet undergone certain modifications, see Chapter 4.

   Cronbach's α is calculated as

$$\alpha = \frac{N^2 \cdot \overline{cov}}{\sum s^2_{item} + \sum cov_{item}}$$

and thus measures the internal consistency of a scale, i.e. the scale reliability (N = # items, $\overline{cov}$ = average covariance between items, s^2 = variance, cov = covariance). Cronbach's α is sensitive towards reverse-coded items (they should be recoded), and the factor N^2 rewards larger numbers of items in a scale, cf. Field (2009). Depending on which data is measured, values above .6 or above .7 are deemed acceptable or good.

   In accordance with other research (Wild, 2000, 2005), the subscales of *Metacognition* do not meet the standards of internal reliability (see Table 5.26), but *Metacognition* as a whole does. This phenomenon stresses the special role of metacognitive learning strategies. They appear difficult to measure (at least when using self-reporting questionnaires), although their importance is generally accepted. Metacognitive aspects are seen as levers with regard

Table 5.26: Cronbach's α for Scales, Data from Three Project Cycles (2011/12 - 2013/14)

Learning Strategy / LIST scale	# items	α	N
Cognitive learning strategies	23	.858	1197
Organizing	8	.814	1605
Elaborating	8	.766	1606
Repeating	7	.726	1646
Metacognitive learning strategies	11	.728	1513
Planning	4	.642	1875
Monitoring	4	.560	1824
Regulating	3	.539	1898
Resource-related learning strategies	35	.802	1000
Effort	8	.757	1556
Attention	6	.749	1663
Time Management	4	.756	1901
Learning Environment	6	.700	1721
Peer Learning	7	.783	1664
Using Reference	4	.765	1920

to influencing future learning behaviour, as they represent the individuals' consciousness, reflections and decisiveness to modulate their learning routines in reference to examination success.

5.6.2 Subsuming aspects of learning behaviour

Exploratory factor analysis (EFA) aims at measuring the hidden constructs behind the data. In most cases of data acquisition, there are numerous items describing different aspects of fewer (more abstract) constructs hidden behind them. In LIST, for example, the items from the *Time Management* scale operationalise a proband's ability to plan and organise their time efficiently. The researcher's main interest is not, however, whether someone writes down timetables, but if they generally make an effort to use their study time properly, so it makes sense to group items. EFA tries to explain the maximum amount of common variance (whereas principal component analysis, PCA, aims at explaining the maximum amount of total variance – which does not really matter in this case) by calculating the loading (projection) of each vector item on an optimised (in the sense of having the smallest deviations) common direction, then named a factor. The higher a loading, the higher the collinearity of the items vector and the common direction. Of course, an item can load inconsistently (crossload), if it is placed in between two different direction vectors. Ideally, EFA will produce a more easily assessable picture out of a large number of items, thus allowing for interpretation.

In order to conduct EFA, the number of factors has to be determined, the kind of rotation must be fixed, and the loadings of the individual items have to be checked. These steps are described now in detail.

Because of the first impression that, in contrast to the theory behind LIST, the data does not consist of 12 factors, the following calculations assume that the LIST questionnaire can be structured in no more than ten factors, as *Metacognition* is regarded as one component. For the exploratory factor analysis, maximum likelihood extractions of factors was chosen, which should allow to later generalise the findings from the sample participants (the first year engineering students present at the time of the survey) to the entire population (first year engineering students), see Field (2009, p. 637). And maximum likelihood extraction is a good choice when the sample stems from a multivariate normal distribution and it additionally provides us with a statistical test for the significance for each factor. Furthermore, pairwise exclusion of cases was opted for (adequate in a sample size as large as the one at hand) because a considerable number of questionnaires was incomplete. Orthogonal as well as oblique rotations were tried, hoping that the problems with the *Metacognition* scale may be reduced under oblique rotation, which does not deny that there may be close interrelations between the scales.

Ideally, all items were to load on only their intended factors with an impact $> .3$, without any crossloadings. Our data did not meet this ideal, its factor structure differed slightly. It was found that the method of rotation does not matter. Both orthogonal (varimax) or oblique (direct oblimin) rotation show nearly identical results. The details are given in Table 5.27.

All items were relatively normally distributed with a tendency to right skew (though all ≤ 1.023 which means less high than low scores) and a vast majority of low negative kurtosis (absolute values ≤ 1.1114, typical of a 4-point Likert scale). The results suggest a problem with multicollinearity, though, which is supported by a look at the factor loadings.

Table 5.27: EFA of LIST with 10 Factors, Pairwise Exclusion of Cases

KMO	.913
KMO values for individual items	$\geq .764$
Bartlett's test of sphericity	$\chi^2(2346) = 37916.098;$ $p = .000$
10 components explain variance	45.65%
Determinant of correlation matrix	$11.15 \cdot 10^{-9}$
Residuals $> .05$	2.0%
Goodness of fit	2.52

In the factor analyses conducted, some items load on factors they were not intended for when the LIST questionnaire was designed. *Planning* items always load on the same factor as *Time Management* – which is not really a problem as they do so consistently, and the topics are clearly related. Some other items proved more difficult and their scales must be interpreted with care in further investigations. These comprise all *Monitoring* and *Regulating* items – which means that the metacognitive items do not form separate scales. This left nine of the original twelve factors. The scree plot (Figure 5.11) suggested between seven and twelve factors, the highest number is supported by the original LIST structure.

In the end, a good scenario was achieved by extracting nine factors using maximum likelihood extraction, direct oblimin rotation and listwise exclusion of cases. The item *attention3* was deleted because of weak and unsystematic loadings. Items *environ2, repeating1* and *repeating4* were kept despite irregular loadings, as their behaviour was judged acceptable. Table I.1 shows the loadings (*Regulating* and *Monitoring* items deleted as well as *attention3*). The nine factors explain 44.23% of variance, KMO = .902, $\chi^2(1830) = 29763.560$, $p = .000$, and the KMO values for individual items are sufficiently large (only two are between .7 and .8, the rest are $> .8$). The goodness of fit index is acceptable (according to Kline, 1994), as the ratio $\chi^2/df = 3275.920/1317 \approx 2.49$ is between 2 and 3. Better fits can be obtained with more factors, but then the factors lack meaningful descriptions.

Tables 5.28 and 5.29 reflect the findings from above (identical calculations were performed for all data from the first until the fourth project cycle, from 2010/11 until 2013/14) – and it yields an insight into the variability that can

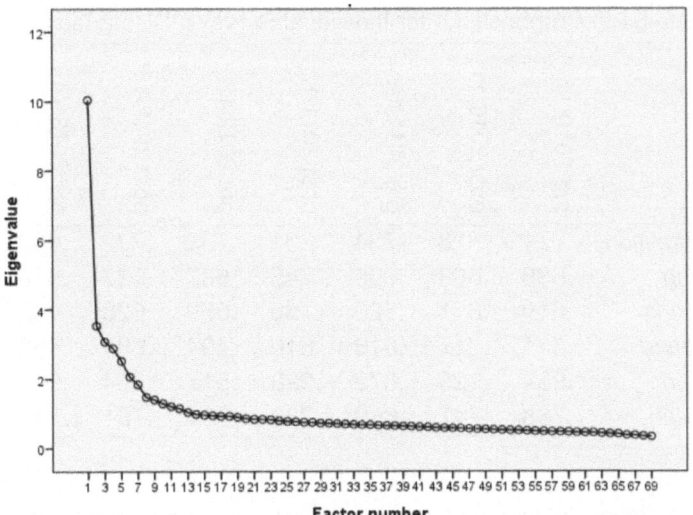

Figure 5.11: Scree Plot for Exploratory Factor Analysis of LIST

Table 5.28: Cronbach's α for Individual Surveys, Modified Scales

	pre 2010/11	post 2010/11	pre 2011/12	post 2011/12	pre 2012/13	post 2012/13	pre 2013/14	post 2013/14
n (with ≥ 60 LIST items)	205	138	345	103	421	254	955	296
	(200)	(134)	(231)	(99)	(376)	(233)	(780)	(253)
Organizing	.861	.902	.808	.852	.820	.805	.762	.851
Elaborating	.808	.835	.748	.795	.775	.770	.703	.787
Repeating	.790	.811	.716	.780	.676	.695	.719	.705
Effort	.793	.787	.713	.844	.748	.805	.740	.731
Attention, att3 deleted	.923	.913	.854	.918	.852	.821	.865	.885
Time Man. / Planning	.805	.807	.745	.817	.801	.801	.771	.824
Learning Environment	.745	.759	.708	.725	.695	.729	.680	.701
Peer Learning	.825	.838	.733	.798	.804	.770	.793	.803
Using Reference	.787	.831	.704	.789	.796	.768	.735	.808

Table 5.29: Cronbach's α for Individual Surveys, Discarded Scales

	pre 2010/11	post 2010/11	pre 2011/12	post 2011/12	pre 2012/13	post 2012/13	pre 2013/14	post 2013/14
Metacognition	.771	.828	.734	.831	.732	.732	.669	.740
Planning	.636	.663	.608	.745	.637	.627	.616	.675
Monitoring	.619	.651	.520	.730	.559	.628	.482	.574
Regulating	.511	.717	.616	.610	.494	.604	.450	.610
Attention	.934	.925	.872	.928	.517	.854	.549	.909
Time Man.	.788	.747	.669	.736	.781	.761	.747	.794

happen to data collection over a period of time and resembles the usual values for LIST, see section 2.2.2 . Again, the subscales of *Metacognition* do not meet the standards of internal reliability, as their Cronbach's α values are under .7 but for three exceptions (*Planning* and *Monitoring* in the post survey from 2011/12, and *Regulating* in the post survey from 2010/11). Also, *Metacognition* as a whole seems more reliable. There are several more α values below the standard of .7, distributed between the scales and years. Deleting *attention3* has greatly improved the two worst alphas, while the others have hardly suffered. This supports the decision to do so.

We see that the pre surveys comprise worse internal reliability than the post surveys. Looking at the participation numbers, there is an obvious reason for that: At the beginning of the semester, there were many more students present in the lecture halls when the questionnaires were distributed. Taking into account that the background project MP^2-Math/Plus had not been presented to them, and that the time needed to complete the questionnaire was often felt to load on free time or breaks, the care needed to read the questions and to reflect on possible answers may not have been exercised as much as would have been necessary to gain reliable data. In the post surveys, the situation was quite different: Those attending the survey (i.e. those present in a specific lecture at the end of the semester) were likely to be more determined and more thorough, which might reflect in a more careful filling in of the LIST questionnaire. What is more, at that time, MP^2-Math/Plus had been part of some students' weekly routines, and may have gained a positive reputation, resulting in an improved attitude towards its representatives and their requests. In the specific situation of a survey, these conditions are also influenced by the lecturers' attitude towards projects disrupting their teaching routines in general.

5.6.3 Exploring the model fit

Once the items are assigned to their respective factors, a confirmatory factor analysis (CFA) was conducted. This tests if the model presented above yields an acceptable fit for the data at hand. The adjusted model (see Appendix I) with shortened *Attention* scale and deleted *Monitoring* and *Regulating*, however, did not submit itself to CFA easily.

CFA allows to restrict the item loadings on the factors they belong to, whereas in EFA, an item can load on more than one factor. The multiple restraints entered into the model (in the form that items are not allowed to load freely on more than one factor) are then tested with the help of fit indices to ascertain if the assignments of items to factors are appropriate.

To judge the development of the model fit in CFA, a comparative fit index (CFI), two badness-of-fit indices (root mean square error of approximation, RMSEA, and standardised root mean residual, SRMR), and two goodness-of-fit indices (Akaike Information Criterion, AIC, and Bayesian Information Criterion, BIC) were calculated to balance for complexity of the model, sample size and robustness against violations of the underlying distribution, (Table 5.30). The RMSEA, measuring an error, is supposed to be small, possibly $< .08$, if possible $< .05$. The SRMR computes residuals and is best when $< .08$. The CFI, on the other hand, compares the model to the data and should therefore be close to 1.

The starting point is the postulated model for LIST, containing 12 factors (*Metacognition* are treated as three factors). This takes into account that for the purposes at hand, the original scale *Critical Checks* was deleted from the beginning, and the items from the scale *Using Reference* were rephrased to include digital ways of procuring data.

As concluded in section 5.6.2, the items from the metacognitive scales as well as *attention3* do not load as expected. Therefore the model was recalculated without *attention3* and without *Monitoring* (second row in Table

Table 5.30: Confirmatory Factor Analysis for LIST Modifications

	$\chi^2(df)$	p	RMSEA	CFI	SRMR	AIC	BIC
a)	11464.87 (2211)	.000	0.0420	0.8001	0.0523	11872.87	−5719.76
b)	9707.602 (1897)	.000	0.0417	0.8181	0.0498	10073.60	−5036.51
c)	10329.8 (1960)	.000	0.0424	0.8089	0.0508	10699.80	−4903.98
d)	1191185 (1734)	.000	0.0497	0.7519	0.7011	12225.85	−1565.37
e)	12417.23 (1743)	.000	0.0508	0.7397	0.6976	12713.23	−1129.95

specifications: a) LIST without modification, b) *attention3* and *Monitoring* scale deleted, c) *attention3* and *Regulating* scale deleted, d) *attention3*, *Monitoring* and *Regulating* scales deleted, factor variances set to 1, e) *attention3*, *Monitoring* and *Regulating* scales deleted, factor variances set to 1, *Planning* and *Time Management* as one factor

5.30) as well as without *attention3* and without *Regulating* (third row in Table 5.30). The model with *attention3*, *Monitoring*, and *Regulating* deleted at the same time did not converge, possibly because of under-identification. If factor variances are set to 1, this problem is solved, though (fourth row in Table 5.30). It now remains to merge *Planning* and *Time Management* (last row in table 5.30) in order to gain an overview of the impact of the modifications, which are visualised in Figure 5.12 (see also Figure 2.9).

The CFI and the RMSEA are not very sensitive towards the modifications, they stay in the required range (although the final values do not satisfy). The SRMR paints another picture, though: It indicates that the modifications worsen the fit of the model, particularly when setting the factor variances to 1.

A look at the Cronbach's α values for individual surveys rounds up the exploration of factors, loadings and internal reliability. The details are given in Tables 5.28 and 5.29, including the first project cycle, treated as a pilot year, in 2010/2011.

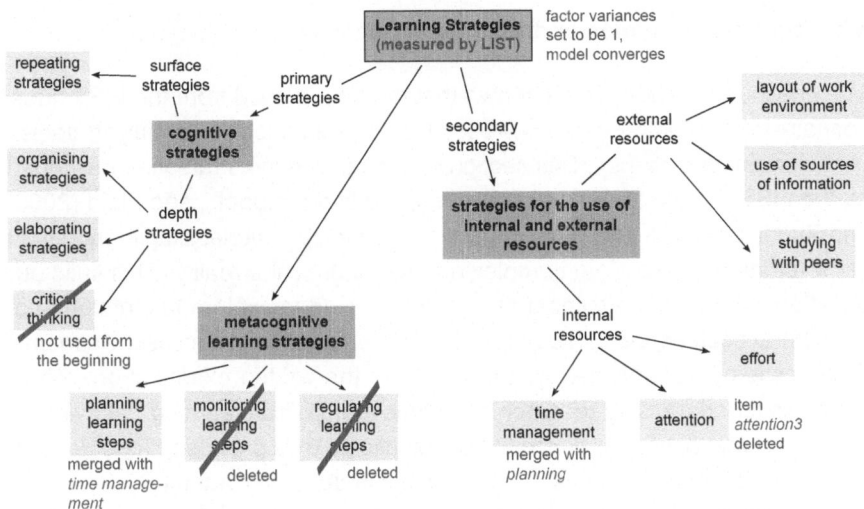

Figure 5.12: Learning Strategies Measured by LIST after Modifications

5.6.4 Summary

The objective of the above calculations was to explore if the data collected for the research at hand with the LIST questionnaire is suitable for describing learning behaviour. It is clear now that this is the case, with one restriction: The metacognitive scales will have to be interpreted with care. The deletion of one item improves the internal reliability of the *Attention* from $\alpha = .749$ to $\alpha = .858$. The shortening of the *Attention* scale by one item as mentioned above, on the other hand, has no big implications on the interpretability of the data. In fact, the item deleted (*I find myself thinking of completely different things*) has no direct implication to the learning situation, and therefore is no heavy loss. It is not the aim here to shorten or otherwise modify the LIST questionnaire, though calculations have shown this to be possible and reasonable (Griese, Lehmann, & Roesken-Winter, 2015).

In sum, the data obtained from the surveys via LIST may be used to compile **empirical arguments** on the differences and the development of learning strategies as described in Figure 5.12, although interpretations should consider some difficulties with the metacognitive scales and model fit.

5.7 Learning Strategies and Academic Success

The previous section 5.6 has shown that the data gained from the LIST questionnaire in MP^2-Math/Plus can be used for the exploration of learning strategies (with certain restrictions). This section connects learning strategies (described by the LIST data) with academic success via linear models, see Field (2009) and Rudolf and Müller (2012). The aim is to develop a model simple enough to allow for interpretation, yet complex enough to describe reality. This suits our situation, where multiple linear regression is used to explore the influence of the different categories of learning behaviour on academic success. *Academic Success* is operationalised by the marks in the written examination. Thus, the basis for addressing research questions four (*Which learning behaviour is connected with academic success?*, see section 3.3) is provided. Together with the other research questions, it is discussed in depth in Chapter 6.

After displaying the data pool and its parameters (section 5.7.1), this section takes a detour and explores the impact of prior skills on the examination result (section 5.7.2). Then learning strategies are investigated as predictors, which needs various calculations and considerations (section 5.7.3) until a final model is found (section 5.7.4) and approved (sections 5.7.5 and 5.7.6). Section 5.7.7 summarises the findings.

5.7.1 Data pool and descriptive statistics

The complete sample from the second to the fourth project cycle[19] was used for the calculations. Only the data from the post surveys was entered into the analysis, as this best describes the learning behaviour crucial for success in the written examination a few weeks later, at the end of the semester. The sample contained both MP^2-Math/Plus participants (121 data sets) and non-participants (533 data sets), but only 351 with matched results for the examination[20]. The relatively small numbers are also due to the fact that the survey was conducted in a break in the mathematics lecture, where attendance is optional.

---

[19] The first project cycle will be taken as a pilot year, for the reasons given in section 5.6.1. For explorations of their learning strategies see (Griese et al., 2011).

[20] This reflects the fact that many students take part in the course but do not sit in on the examination, see also section 5.5.2.

The scale scores were calculated by the affine linear transformation

$$\frac{100}{3} \cdot \left(\frac{1}{n} \sum_{i=1}^{n} x_i - 1 \right) = \frac{100}{3} \cdot (\bar{x} - 1),$$

where n is the number of items in a scale, and $x_1, \ldots, x_n$ are their scores[21]. This takes into account that there are different numbers of items in particular scales, and it yields values from 0 to 100. 0 stands for the lowest possible value (where all items in the scale are marked with 1 for *very seldom* in a Likert scale), and 100 means all items in this scale were marked with 4 for *very often*. The descriptive statistics for all LIST scales (including the original LIST, the original LIST with *Metacognition* as one scale, and the modified LIST) are specified in Table 5.31. The lowest scores ($\approx$ 40, "seldom") are found in *Repeating* and *Time Management*, the highest ($>$ 60, "often") in *Regulating*, *Learning Environment* and *Using Reference*.

The scale scores are used as predictors to calculate their influence on the outcome variable, *Academic Success*, represented by the mark in the written examination[22]. Constants, coefficients b, their standard errors, standardised coefficients β, their significance values, R^2 and ΔR^2 are computed, like in Griese and Kallweit (2016) for a similar setting. Missing data is eliminated pairwise.

Linear models are calculated while using three different ways of entering or removing predictors: backward, forward, and stepwise, see Rudolf and Müller (2012). When choosing backward removal of predictors, all predictors are initially entered into the model, and single predictors are removed according to their contribution to the model. It is tested if the change of the F-value is significant, as a standard to the level of .10. The forward method starts with the predictor that correlates highest with the outcome variable and enters new predictors as long as the change of the F-value is significant to the level of

---

[21] When calculating linear models, the mean value of a score would produce identical βs and significances, but a different constant and different b values with different standard errors. The choice for the linear transformation is justified by easier accessibility of the 0 to 100 scale, which is also used in section 5.8. The difference in range between the predictors (between 0 and 100) and the outcome variable (between 0.7 and 5.0) yields relatively small b values, though.

[22] In Germany, 1 and 2 are excellent respectively good marks, 3 is satisfactory, 4 is a pass mark, and 5 means fail. Thus, a negative coefficient is interpreted as an often-used learning strategy entailing a better mark. This does not apply to *Attention*, however, as all items in this scale are reverse-coded.

Table 5.31: Descriptive Statistics for LIST Scales,
Post Surveys from the Second to Fourth Project Cycle

Variable / Scale	Mean	SD	n
Organizing	47.434	22.248	492
Elaborating	48.392	19.019	482
Repeating	40.282	18.007	514
Planning	48.672	22.214	571
Monitoring	46.769	20.850	539
Regulating	65.689	20.516	579
Metacognition	52.814	16.243	483
Effort	58.244	18.942	468
Attention	47.049	24.428	523
Time Management	39.088	23.399	559
Time Management / Planning	43.640	20.265	522
Learning Environment	64.941	18.377	512
Peer Learning	55.527	20.593	492
Using Reference	69.449	23.444	566

Score meanings: [0; 16.67] very seldom,]16.67; 50] seldom,
]50; 83.33] often,]83.33; 100] very often

.05. These significance levels can be adjusted, if the results do not satisfy. The stepwise method combines the backward and forward method.

5.7.2 Appraising prior skills

The variable *Mini Test* reflects academic accomplishment at the beginning of the semester, before any (or much) of the learning strategies that are the focus of MP^2-Math/Plus have been imparted. It measures the score in the first mini exam in percentage of achievement points. This variable correlates with the other variables as shown in Table 5.32. *Mini Test* correlates strongly ($r = -.643$, which means $R^2 > .4$ for this one predictor alone) and highly significantly ($p = .000$) with *Academic Success*. Therefore it is to be expected that whenever *Mini Test* is entered into a linear model, it will outscore the other possible predictors. In effect, when using forward or stepwise methods, or even the backward method with adjusted significance levels of for entry (.005) and removal (.010), a linear model containing *Mini Test* as predictor and outcome variable *Academic Success* has only one other predictor, *Effort*, see Table 5.33.

Table 5.32: Pearson's *r* of *Mini Test* to Other Variables,
$^* < .05, ^{**} < .01, ^{***} < .001$

Academic Success	-.643***	*Attention*	-.198***
Organizing	-.054	*Learning Environment*	.070
Elaborating	-.022	*Peer Learning*	.006
Repeating	-.145**	*Using Reference*	-.035
Effort	.169***	*Time Man. / Planning*	-.056

However, the focus here is not on exploring how previously able students consequently excel in eventually following examinations, but to investigate what part learning behaviour plays for mediocre or weak students when trying to improve their academic achievement. Thus, *Mini Test* will not be entered into any further models – but it is relevant to keep in mind how much variance it can explain.

Table 5.33: Linear Model Containing *Mini Test* as Predictor,
$R^2 = .421$, Durbin-Watson 1.952

Predictor	*b*	SE for *b*	β	Sig.
(Constant)	5.253	0.240		.000
Mini Test	-3.402	0,265	-.611	.000
Effort	-.011	0.004	-.138	.004

Table 5.34: Pearson's r for Modified LIST Scales and Academic Success, * < .05, ** < .01, *** < .001

	Organizing	Elaborating	Repeating	Effort	Attention	Environment	Peer Learning	Reference	Time M. / Planning
Academic Success	.031	.005	-.015	-.241***	.235***	-.049	.046	.032	-.024
Organizing		.290***	.603***	.451***	-.192***	.331***	.239***	.261***	.517***
Elaborating			.320***	.338***	-.153***	.139**	.282***	.257***	.300***
Repeating				.524***	-.205***	.251***	.293***	.234***	.500***
Effort					-.531***	.455***	.253***	.389***	.444***
Attention						-.303***	-.136**	-.218***	-.167***
Environment							.077	.288***	.332***
Peer Learning								.118**	.208***
Using Reference									.192***

5.7.3 Testing different linear models

First, the correlations between the LIST scales and *Academic Success* were considered by calculating Pearson's correlation coefficients. The results for the modified LIST scales (see section 5.6) are shown in Table 5.34. *Academic Success* correlates highest with *Effort* ($r = -.241$, $p = .000$) and *Attention*[23] ($r = .235$, $p = .000$), though the correlations are rather weak. In both cases, the correlations are highly significant, though, and work in the expected direction, namely that more effort / exertion and a more attention / concentration are connected with better marks in the examination.

The weakness of the correlations in Table 5.34 may be caused by the fact that learning behaviour does not account for intellectual capacity, understanding, or even foreknowledge, which play a role in *Academic Success*. The special role played by *Effort* also shows in its correlations with the other scales, see Table 5.34. There are five other scales that correlate both considerably ($r > .4$) and significantly ($p < .001$) with *Effort*: *Organizing* ($r = .451$, $p = .000$), *Repeating* ($r = .524$, $p = .000$), *Attention* ($r = -.531$, $p = .000$), *Learning Environment* ($r = .455$, $p = .000$), and *Time Management / Planning* ($r = .444$, $p = .000$). This makes *Effort* a candidate for a mediator[24] when compiling a linear model with the outcome variable *Academic Success*.

Then, in order to explore what variance of *Academic Success* can be explained by the variables at our disposal, first the linear models with twelve (original LIST) respectively nine (modified LIST, see section 5.6) predictors are calculated, see Tables 5.35 and 5.36.

Both models show a highly significant ($p = .000$) and strong ($\beta = -.475$ respectively $\beta = -.523$) influence of *Effort* (in the sense that high effort entails better marks in the examination) and a significant ($p \leq .002$) though weaker ($\beta = .255$ respectively $\beta = .240$) influence of *Using Reference*, although in the sense that more frequent use of works of reference or the Internet entails worse marks. The predictors *Time Management*, *Planning* or the combination of both yield considerable influence ($\beta = .180$ for *Planning* and $\beta = .229$ for *Time Management / Planning*), though it is not always significant. In both cases,

---

[23] Please note that *Attention* is reverse coded.

[24] A mediator variable is a predictor that, when entered into a linear model, increases the influence of another predictor variable on the outcome variable.

Table 5.35: Regression Model with Twelve Predictors and Outcome Variable
  Academic Success, $n = 157$, $R^2 = .28$,
  $* < .05$, $** < .01$, $*** < .001$

Predictor	b	SE for b	β	Sig.
(Constant)	2.972	0.724		.000
Organizing	0.003	0.006	.053	.585
Elaborating	0.011	0.007	.152	.086
Repeating	0.003	0.009	.035	.743
Planning	0.012	0.007	.180	.072
Monitoring	-0.012	0.007	-.173	.086
Regulating	-0.012	0.007	-.156	.081
Effort	-0.037	0.009	-.475	.000***
Attention	0.006	0.006	.090	.308
Time Management	0.005	0.005	.085	.363
Learning Environment	0.004	0.007	.046	.595
Peer Learning	0.007	0.006	.094	.225
Using Reference	0.016	0.005	.255	.001**

more time planning entails worse marks. The highest R^2 observed is .28, which can be regarded as an upper bound for all models with less predictors.

The calculations for a linear model with the outcome variable *Academic Success* were made with the original LIST scales and with the modified LIST scales (see section 5.6). Only the backward method with the original LIST scales yielded a model with seven scales, all other methods and scales produced the model with four predictors shown in Table 5.37. Again, *Effort* stands out; it has the highest (and significant) β. The negative algebraic sign means the influence is as expected. That applies to *Attention*, too, in spite of the positive sign, because this scale is reverse-coded. The other two predictors, *Organizing* and *Using Reference*, apparently contribute to *Academic Success* in an unexpected way: It seems that the more a student re-organises the subject material, and the more he or she refers to books, a script or the Internet, the worse he or she scores in the written examination. This interpretation can be rejected, though, on the grounds that neither *Organizing* nor *Using Reference* correlate

Table 5.36: Regression Model with Nine Predictors and Outcome Variable
Academic Success, $n = 163$, $R^2 = .24$,
$* < .05$, $** < .01$, $*** < .001$

Predictor	b	SE for b	β	Sig.
(Constant)	2.499	0.688		.000
Organizing	0.003	0.006	.041	.661
Elaborating	0.005	0.006	.067	.392
Repeating	-0.004	0.008	-.046	.642
Time Man. / Planning	0.017	0.006	.229	.010*
Effort	-0.040	0.008	-.523	.000***
Attention	0.006	0.005	.092	.289
Learning Environment	0.004	0.007	.053	.530
Peer Learning	0.007	0.006	.086	.266
Using Reference	0.015	0.005	.240	.002**

significantly with *Academic Success*, see Table 5.34. This phenomenon is
called suppression: *Organizing* and *Using Reference* are not correlated to
Academic Success, but they appear to make significant contributions in the
linear model. This happens when a predictor variable (here *Organizing* or *Using
Reference*) suppresses unwanted variance in another predictor (here *Effort*).

5.7.4 Inspecting the final model

The calculations so far leave two remaining predictors, which result in the linear
model presented in Table 5.38, whose explanation of variance ($R^2 = .074$) is
modest, though. The results suggest collinearity between *Effect* and *Attention*,
meaning that both variables describe related concepts and one might therefore
be redundant in a linear model. The test for multicollinearity[25] delivers a
tolerance of .718, and the variance inflation factor (VIF), the reciprocal value
of tolerance, of 1.392. Both imply that there is no collinearity between these

---

[25] For this test, a multiple linear regression is computed, where one predictor becomes the outcome
variable. If one predictor can be used to predict another, this implies collinearity, resulting
in a high R^2, and therefore in a small tolerance (=1 − R^2). A tolerance < .1 is considered
conspicuous, see Rudolf and Müller (2012, p. 76).

Table 5.37: Regression Model with Four Predictors and Outcome Variable
Academic Success, $n = 266$, $R^2 = .111$, Durbin-Watson 1.926

Predictor	b	SE for b	β	Sig.
(Constant)	2.863	0.488		.000
Effort	-0.022	0.006	-.287	.000
Organizing	0.010	0.004	.153	.021
Attention	0.009	0.004	.141	.041
Using Reference	0.008	0.004	.134	.035

two predictors that might compromise the linear model. The Durbin-Watson
statistic of 1.946 is close enough to its ideal value of 2 to dissipate concern on
autocorrelation.

Table 5.38: Final Regression Model with Two Predictors and Outcome Variable
Academic Success, $n = 268$, $R^2 = .074$, Durbin-Watson 1.946

Predictor	b	SE for b	β	Sig.
(Constant)	3.340	0.462		.000
Effort	-0.013	0.005	-.162	.021
Attention	0.009	0.004	.149	.033

The final model with the two predictors meets the usual standards because
when looking at the scatter plot of the standardised predictors and the standardised residuals (Figure 5.13), there is no indication to reject the assumption
of homoescadicity, as there is no flare shape typical of heteroescadicity. The
arrangement of the scattered dots along declining lines is due to the fact that the
variable *Academic Success* does not assume all values between its minimum
0.7 and its maximum 5.0, but only those representing German marks. Neither
Effort nor *Attention* pass the Kolmogorov-Smirnov or the Shapiro-Wilks tests
for normal distribution (which would be ideal), but for *Effort*, the deviation is in
acceptable limits, see Figure 5.14.

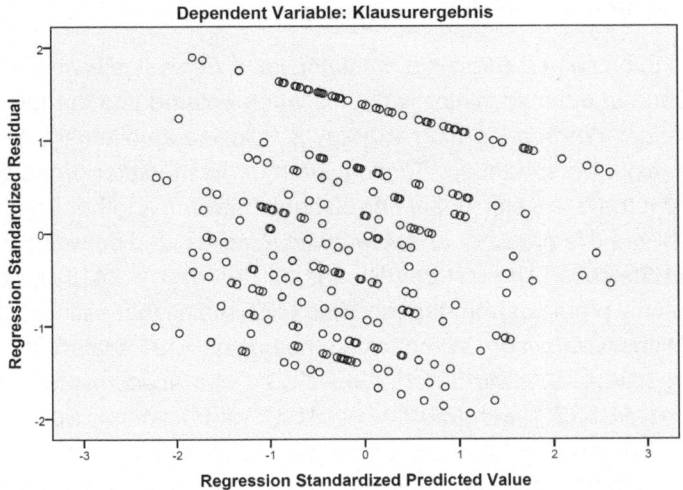

Figure 5.13: Scatterplot for the Linear Model with Two Predictors, see Table 5.38

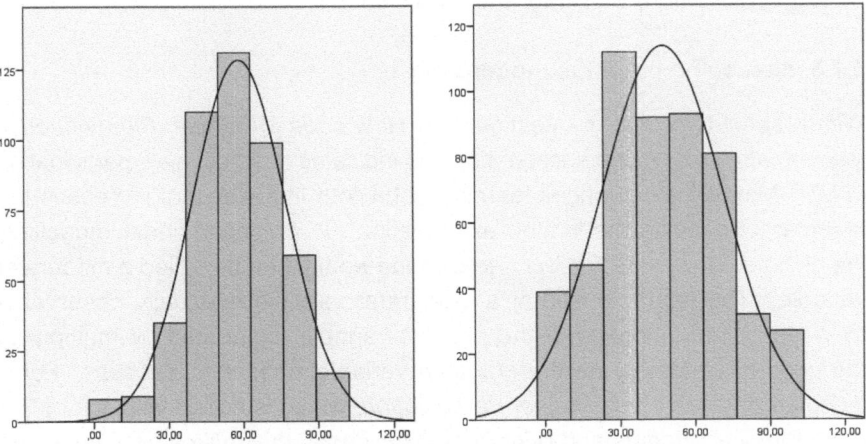

Figure 5.14: Histograms for *Effort* (left) and *Attention* (right) with Normal Distribution Curves

5.7.5 Screening for possible mediators

It remains to be seen if *Effort* is a mediator, i.e. if *Effort* significantly increases the influence of other predictor variables when entered into the model. This makes sense: When a learning strategy is followed ambitiously, its positive influence might be enhanced. Testing *Effort* for its mediator property is only scheduled if there are significant correlations between another predictor and *Effort*, between this predictor and *Academic Success*, and between *Effort* and *Academic Success*. The last condition is fulfilled ($r = -.241^{***}$), see Table 5.34. The only predictor from the modified LIST scales that satisfies the other two conditions is *Attention*. When taking the original LIST scales into account, *Monitoring* ($r = .479^{***}$ with *Effort*, $r = -.163^{**}$ with *Academic Success*) and *Regulating* ($r = .532^{***}$ with *Effort*, $r = -.163^{**}$ with *Academic Success*) are to be considered, too. A comparison of the linear model with *Effort* and *Attention* as predictors to that with only *Attention* shows that *Effort* is not a mediator, as its presence in the model decreases *Attention*'s β value from $\beta = .235^{***}$ to $\beta = .149^*$. The same is true for *Monitoring*, whose β falls from $-.163^{**}$ (with *Effort*) to $-.062$ (not significant), and for *Regulating*, whose β dwindles from $-.163^{**}$ (with *Effort*) to $-.049$ (not significant).

5.7.6 Assessing potential moderators

With respect to research question five (*How does MP^2-Math/Plus influence learning strategies?*, see section 3.3), it is indicated to scrutinise if participation in MP^2-Math/Plus influences learning behaviour in a way that increases the chances of passing the written examination. In (multiple) linear modelling, the dichotome variable *Project Participation* would then be called a moderator variable. The characteristic of a moderator variable is usually observable by a significant influence of the product variable (computed by multiplying the centred variables) on the outcome variable *Academic Success*. For a dichotome variable like *Project Participation*, which is coded with 1 for MP^2-Math/Plus participants and 0 for non-participants, this makes no sense, and we will therefore compute separate models for MP^2-Math/Plus participants and non-participants, both with the predictors *Effort* and *Attention* and the outcome variable *Academic Success*. The ensuing linear models are shown in Table 5.39.

Table 5.39: Linear Models Testing MP^2-Math/Plus Project Participation for Moderator Property

Predictor	b	SE for b	β	Sig.
MP^2-Math/Plus participants, R^2 = .020				
(Constant)	3.617	0.850		.000
Effort	-0.009	0.010	-.139	.330
Attention	0.000	0.008	.003	.982
Non-participants, R^2 = .095				
(Constant)	3.223	0.559		.000
Effort	-0.013	0.007	-.160	.050
Attention	0.012	0.005	.190	.020

The results are weakened by the low numbers of data sets that these restrictions invariably involve. When testing MP^2-Math/Plus project participation for its moderator property, only 66 data sets of participants (and 197 of non-participants) are left. Nevertheless, the contribution of *Attention*, which measures the amount of distraction and procrastination, vanishes for the project group, and loses its significance.

5.7.7 Summary

In the final model, only *Effort* and *Attention* are included as exerting influence on *Academic Success*. They were found to be not collinear, meaning they mirror different aspects of learning behaviour. Other learning strategies were removed from the model due to insufficient correlation with the outcome variable. *Effort* was found to be no mediator, i.e. frequent reporting of *Effort* does not enhance the influence of other learning behaviour. Participation in MP^2-Math/Plus did not moderate the impact of learning behaviour in a significant way, although in the final model for project participants, the influence of *Attention* has vanished and lost its significance.

5.8 Development of Learning Strategies

This section investigates the development of the learning strategies described by our LIST data, i.e. the changes between the pre survey at the beginning of the semester and the post survey at the end. The pre surveys were conducted a few weeks after the lectures had started, when students had had some experience of university life and learning, but no tests or examinations had yet taken place. The post surveys were carried out in the last or second last week of lectures, when the students had lived through their first university semester, and the final examinations were imminent.

The differences between the pre and post scores can therefore describe the development of learning behaviour undergone by first-year engineering students. As section 2.1 has illustrated the differences between school and university mathematics, it is to be expected that the cohort as a whole will show some development because students adapt to the demands of tertiary mathematics. So this section particularly explores if MP^2-Math/Plus participants' adaptations to university learning behaviour differ from those of non-participants, and for which learning strategies this can be observed. This forms the basis for discussing research question five, *How does MP^2-Math/Plus influence learning strategies (and motivation)?*, see section 3.3, which can be found in Chapter 6. As section 5.7 has revealed the importance of *Effort* and *Attention*, we will be paying special attention to these scales. Section 5.8.4 summarises the findings.

5.8.1 Data characteristics and properties

The data from the second to fourth project cycle[26] was used to investigate changes in learning behaviour. The descriptive statistics are given in Table 5.40[27] for the modified LIST scales (see section 5.6), which have proven useful in section 5.7. Most scales show less frequent use of the respective learning behaviour from the pre survey to the post survey, most clearly *Elaborating* and *Repeating*. *Attention*, whose items are reverse-coded, shows a distinct decrease of distraction, i.e. increase of concentration from pre to post. *Time*

---

[26] As before, the first project cycle was treated as a pilot year.
[27] The statistics for the post scales are identical with those from Table 5.31. They are repeated here to enable comparisons.

Management / Planning and *Learning Environment* show a slight increase whose significance is doubtful, though.

Table 5.40: Descriptive Statistics for Pre and Post Scales

Variable / Scale	Mean	SD	n
Organizing, pre	50.68	19.99	1113
Organizing, post	47.43	22.25	492
Elaborating, pre	57.05	16.64	1124
Elaborating, post	48.39	19.02	482
Repeating, pre	47.36	18.05	1132
Repeating, post	40.28	18.01	514
Effort, pre	60.62	16.87	1088
Effort, post	58.24	18.94	468
Attention, pre	54.87	23.51	1200
Attention, post	47.05	24.43	523
Time Man. / Planning, pre	41.46	18.69	1172
Time Man. / Planning, post	43.64	20.27	522
Learning Environment, pre	60.33	18.83	1209
Learning Environment, post	64.94	18.38	512
Peer Learning, pre	56.10	19.65	1172
Peer Learning, post	55.53	20.59	492
Using Reference, pre	73.52	20.85	1354
Using Reference, post	69.45	23.44	566

There are many more data sets in the pre surveys (in average around 1200) than in the post surveys (around 500). This may cause unwanted bias, so, as we intend to compare the pre and post scores, consequently, only paired data sets are entered into further analyses. This way individual high or low scores are accounted for, and the reference groups are identical for the pre and the post data. The numbers of paired data sets for the different project years are given in Table 5.41, resulting in a total of $n = 255$. The descriptive statistics for the data sets that can be paired are presented in Table 5.42, Figure 5.15 visualises them. The scores are slightly higher than those from Table 5.40,

Table 5.41: Number of Paired Data Sets

Project Cycle	Second	Third	Fourth	Sum
# Paired Data Sets	24	133	98	255

particularly the ones from the pre survey, with the exception of *Attention, post*[28], which stands for less distraction among the students whose data could be paired.

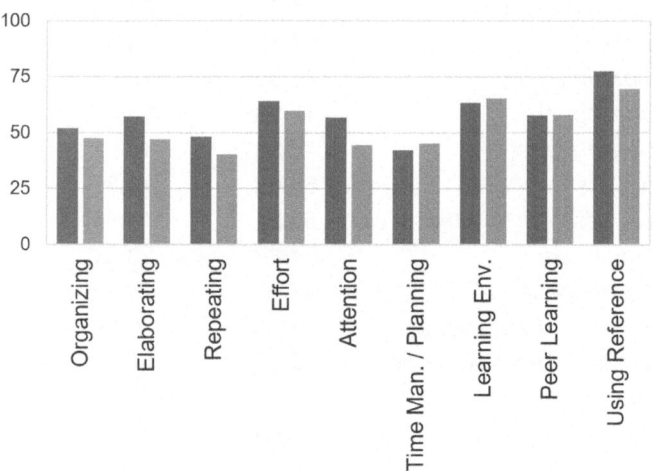

Figure 5.15: Diagram for Table 5.42,
blue / left column = pre survey, green / right column = post survey

All variables describing differences between the pre and the post scores are calculated via

$$\text{score}_{pre} - \text{score}_{post} = \text{difference}.$$

This way, a positive value of a difference variable describes the scenario that a learning strategy has a higher score in the pre survey, i.e. that it was allegedly employed more often at the beginning of the semester than at the end. We see that for the paired data sets, as before, the means usually decrease from pre

---

[28] All items in the *Attention* scale are reverse-coded and describe diversion rather than concentration.

to post survey. This means that on the whole, students reported less frequent use of the learning strategies shortly before the examinations.

Table 5.42: Descriptive Statistics for Pre and Post Scales
            Computed from Paired Data

Variable / Scale	Mean	SD	n
Organizing, pre	51.93	20.97	203
Organizing, post	47.64	23.36	207
Elaborating, pre	57.22	16.98	202
Elaborating, post	47.19	18.97	197
Repeating, pre	48.13	18.51	204
Repeating, post	40.15	17.58	216
Effort, pre	64.29	17.20	203
Effort, post	59.79	19.80	200
Attention, pre	56.83	26.24	210
Attention, post	44.36	23.99	220
Time Man. / Planning, pre	42.08	18.51	210
Time Man. / Planning, post	45.06	20.49	221
Learning Environment, pre	63.36	19.72	210
Learning Environment, post	65.43	18.57	216
Peer Learning, pre	57.83	19.64	216
Peer Learning, post	58.01	19.15	201
Using Reference, pre	77.61	18.96	230
Using Reference, post	69.58	24.03	232

For comparing the pre and post scores, some assumptions need to be checked when considering a dependent t-test (see Field, 2009). The differences between the scores should be normally distributed, and the data should be measured at interval level. The second condition is fulfilled, as the scale scores will be entered into the calculations. Normal distribution of differences is tested using the Shapiro-Wilk test[29] for normal distribution (see Table 5.43), and

---

[29] The Shapiro-Wilk test works well for sample sizes under 50 and can be used for up to 2,000, see Field (2009).

values for skewness and kurtosis[30] (see Table 5.44). Missing data is eliminated pairwise.

Table 5.43: Shapiro-Wilk Test for Normal Distribution of Pre-Post Scale Score Differences for Paired Data, *n* = 255

Variable / Scale	Statistic	df	Sig.
Diff Organizing	.990	171	.311
Diff Elaborating	.989	161	.255
Diff Repeating	.982	174	.025
Diff Effort	.986	164	.092
Diff Attention	.988	189	.108
Diff Time Management / Planning	.988	185	.119
Diff Learning Environment	.977	183	.004
Diff Peer Learning	.987	175	.115
Diff Using Reference	.972	213	.000

If Sig. > .05 we can reject the alternative hypothesis and assume that the variable is normally distributed.

According to the Shapiro-Wilk test, normality can be assumed for *Diff Organizing*, *Diff Elaborating*, *Diff Repeating*, *Diff Effort*, *Diff Attention*, *Diff Time Management / Planning*, and for *Diff Peer Learning*. When standardising the values for skewness and kurtosis (by dividing them by their standard error), normality can be assumed for all variables except *Diff Learning Environment* and *Diff Using Reference*. Taken together, this leaves three scales for which the supposition of normality might be violated.

How far the distributions of *Diff Repeating*, *Diff Learning Environment* and *Diff Using Reference* are different from normality can be estimated by the use of histograms, stem-and-leaf plots, or QQ-plots. For comparison, the histograms with normal distribution curves are displayed in Figure J.1 together with the histogram of the variable *Diff Organizing* which was rated as almost certainly normally distributed by the above calculations. The conclusion is that the

---

[30] Positive kurtosis means a peak narrower than in a normal distribution, negative kurtosis describes a wider peak. Skewness refers to the symmetry of the distribution; negative skewness means a heaviness to the left side.

Table 5.44: Skewness and Kurtosis for Pre-Post Scale Score Differences
for Paired Data, n = 255

Variable / Scale	Skewness	SE Skewn.	Kurtosis	SE Kurt.
Diff Organizing	0.065	0.186	0.123	0.369
Diff Elaborating	0.011	0.191	-0.246	0.380
Diff Repeating	0.069	0.184	0.693	0.366
Diff Effort	0.062	0.190	-0.357	0.377
Diff Attention	-0.169	0.177	-0.323	0.352
Diff Time Man. / Planning	-0.057	0.179	0.593	0.355
Diff Learning Environment	-0.090	0.180	1.413	0.357
Diff Peer Learning	0.045	0.184	-0.177	0.365
Diff Using Reference	0.459	0.167	0.507	0.332

Standardised skewness and kurtosis values should be close to 0 for normally distributed variables. The range between -2 and 2 is reckoned acceptable (Field, 2009, p. 139).

three seemingly problematic variables do not deviate too much from normality; actually, the histogram for Diff Organizing shows the biggest nonconformity.

The QQ-plots[31] are displayed for the same scales, see Figure J.2. They leave some doubt to the normality of distribution of Diff Repeating and Diff Using Reference. Thus, when comparing pre and post scores for these scales, non-parametric tests are indicated. Otherwise, dependent t-tests are appropriate. According to Table 5.42, we may expect significant changes from pre to post for these two scales. For Learning Environment, the difference of the pre and post mean is so small as to eliminate the necessity of a comparison test as it is.

5.8.2 Pre-post comparisons

Now that the preparations for comparing pre and post scores are complete, differences in the use of learning strategies are the focus of further explorations. The descriptive statistics for the pre and post scales (see Table 5.42) have

---

[31] Ideally, the markings in the QQ-plots should not stray from the line, at least not systematically.

already been observed and revealed the fact that most developments imply less frequent use of learning strategies at the end of the semester. It remains to be seen if these changes are significant, and if there are differences between MP^2-Math/Plus participants and non-participants. The results from the dependent t-tests are presented in Table 5.45, with effect sizes $r = \sqrt{\frac{t^2}{t^2+df}}$. These tests are dispensable for *Repeating*, *Learning Environment* and *Using Reference*, as the assumption of normal distribution is probably violated for them, but for reasons of completeness they are listed here, too.

The highest (and significant) changes can be observed for *Attention* (mean difference = 12.17***, r = .27) and *Elaborating* (mean difference = 11.44***, r = .57) with medium and large effect sizes. Please note that *Attention* is reverse-coded, therefore the positive mean difference here indicates that students reported less distraction in the post survey. For *Elaborating*, the positive mean difference translates into less frequent use of learning behaviour associated with finding connections, practical applications, examples, and visualisation. For the second highest mean differences, *Repeating* (mean difference = 8.81***, r = .43) and *Using Reference* (mean difference = 8.18***, r = .32) we will be checking these scales with a non-parametric test. *Effort*, which has shown its relevance in section 5.7, yields a mean difference of 5.23*** with a medium effect size of r = .34.

When comparing scores from the same persons (as we do with the paired data) where the assumption of normal distribution is violated, the recommended non-parametric test is the Wilcoxon signed-rank test (Field, 2009, p. 552f.). As a rank test, it orders the scores and assigns rank levels. The test then uses the ranks rather than the actual scores for its calculations. As such the Wilcoxon signed-rank test appears weaker than its parametric counterpart, but it is the only option left with the non-normally distributed scales *Repeating*, *Learning Environment*, and *Using Reference*. Particularly *Repeating* and *Using Reference* are of interest as they have shown considerable differences in their means, see Table 5.45. The results from the Wilcoxon signed-rank tests are presented in Table 5.46, with their effect sizes[32] $r = \frac{z}{\sqrt{N}}$. Again, all scales are listed for reasons of completeness, although there already are results for the

---

[32] z is the test statistic; N denotes the number of observations, which is twice the number of cases in our pre-post scenario.

normally distributed ones. As is usual when reporting the results of rank tests, the median values are given, too.

The results support the findings from the dependent t-tests from Table 5.45, and add the insight that for *Repeating* (median$_{pre}$ = 47.62, median$_{post}$ = 42.86, $z = -5.944$***, $r = -.32$) and *Using Reference* (median$_{pre}$ = 83.33, median$_{post}$ = 66.67, $z = -4.541$***, $r = -.22$), the changes from pre to post surveys can be regarded as significant, with medium respectively small effect sizes. For *Learning Environment*, the development from pre to post survey is assessed as not significant.

Table 5.45: Dependent *t*-Tests for Paired Data

Scale	Mean Diff.	SE Mean	t	df	Sig. (2-t.)	Effect Size *r*
Organizing	4.17	1.48	2.811	170	.006**	.21
Elaborating	11.44	1.30	8.773	160	.000***	.57
Repeating [a]	8.81	1.41	6.242	173	.000***	.43
Effort	5.23	1.12	4.676	163	.000***	.34
Attention	12.17	3.23	3.770	188	.000***	.27
Time Management / Planning	-2.34	1.36	-1.717	184	.088	.13
Learning Environment [a]	-1.09	1.44	-0.759	182	.449	.06
Peer Learning	1.03	1.47	0.705	174	.482	.05
Using Reference [a]	8.18	1.65	4.942	212	.000 ***	.32

* $< .05$, ** $< .01$, *** $< .001$; [a] assumption of normal distribution probably violated

Table 5.46: Wilcoxon Signed-Rank Tests for Paired Data

Variable / Scale	Median pre	Median post	z	Sig. (2-tailed)	Effect Size r
Organizing	66.67	50.00	-2.609	.009**	-.14
Elaborating	58.33	45.83	-7.332	.000***	-.41
Repeating	47.62	42.86	-5.944	.000***	-.32
Effort	66.67	62.50	-4.270	.000***	-.24
Attention	60.00	40.00	-3.708	.000***	-.19
Time Management / Planning	41.67	45.83	-1.626	.104	-.08
Learning Environment	61.11	66.67	-0.832	.405	-.04
Peer Learning	57.14	61.90	-0.590	.555	-.03
Using Reference	83.33	66.67	-4.541	.000***	-.22

$* < .05, ** < .01, *** < .001$

5.8.3 Influence of MP²-Math/Plus project participation

This next section takes a look at the development of the use of different learning strategies for MP²-Math/Plus project participants in contrast to non-participants. The comparisons can show if participants, who had a network of interconnected project procedures at their disposal, show a development of learning behaviour different from that of the non-participants. The procedure is the same as in the previous section: Dependent *t*-tests as well as Wilcoxon signed-rank tests are run for all scales, this time separately for MP²-Math/Plus project participants and non-participants. The results can be found in Table 5.47 and in Table 5.48[33].

In all scales but *Repeating* and *Time Management / Planning*, the mean score differences are higher for non-participants, meaning that the tendency to show a decrease in the use of a learning strategy was more pronounced in that group. In other words: for participants of MP²-Math/Plus, the tendency of sinking learning behaviour is reported less often; they relate more active learning behaviour, compared to non-participants. For *Time Management / Planning*, MP²-Math/Plus participants display a tendency opposite to that of non-participants. They report a significant increase in the use of this learning strategy (mean difference = -4.84, *SE* = 2.23,) $t(36) = -2.173$, $p = .036^*$, $r = .34$, whereas non-participants (mean difference = -1.76, *SE* = 1.62) do not significantly change their learning behaviour, $t(146) = -1.084$, $p = .280$.

Concerning *Effort*, which has shown its importance in section 5.7, MP²-Math/Plus participants report to be making the same effort at the end of the semester as at the beginning (mean difference = 2.46, *SE* = 2.35), $t(38) = 1.046$, $p = .302$. Non-participants report a significant decrease in *Effort* (mean difference = 5.78, *SE* = 1.24) with a medium effect, $t(123) = 4.674$, $p = .000^{***}$, $r = .39$. For *Attention*, the situation presents itself thus: The scores decrease less here for MP²-Math/Plus participants (mean difference = 8.73, *SE* = 6.78, $t(41) = 1.288$, $p = .205$) than for non-participants (mean difference = 13.56, *SE* = 3.68, $t(145) = 3.684$, $p = .000^{***}$, $r = .29$), this means participants report less decrease of distraction than non-participants.

---

[33] On the whole, the significance of the test mostly suffers for the group of MP²-Math/Plus participants, due to the smaller numbers.

Table 5.47: Dependent t-Tests for Paired Data, Separately for MP²-Math/Plus Participants and Non-Participants

Scale	Mean Diff.	SE Mean	t	df	Sig. (2-t.)	Effect Size r
Organizing participants	1.21	2.91	0.414	37	.681	.07
Organizing non-participants	4.77	1.71	2.791	131	.006**	.24
Elaborating participants	10.53	3.05	3.449	35	.001***	.50
Elaborating non-participants	11.36	1.41	8.073	123	.000***	.59
Repeating [a] participants	8.99	3.13	2.871	35	.007**	.44
Repeating [a] non-participants	8.72	1.60	5.458	136	.000***	.42
Effort participants	2.46	2.35	1.046	38	.302	.17
Effort non-participants	5.78	1.24	4.674	123	.000***	.39
Attention participants	8.73	6.78	1.288	41	.205	.20
Attention non-participants	13.56	3.68	3.684	145	.000***	.29
Time Man. / Planning participants	-4.84	2.23	-2.173	36	.036*	.34
Time Man. / Planning non-part.	-1.76	1.62	-1.084	146	.280	.09
Learning Environment [a] participants	-0.75	3.02	-0.248	36	.805	.04
Learning Environment [a] non-part.	-1.53	1.61	-0.951	144	.343	.08
Peer Learning participants	-1.36	2.85	-0.477	41	.636	.07
Peer Learning non-participants	-1.77	1.72	1.027	131	.306	.09
Using Reference [a] participants	3.46	3.22	1.075	52	.287	.15
Using Reference [a] non-participants	9.75	1.93	5.053	158	.000***	.37

* $< .05$, ** $< .01$, *** $< .001$; [a] assumption of normal distribution probably violated

Table 5.48: Wilcoxon Signed-Rank Tests for Paired Data, Separately for MP^2-Math/Plus Participants and Non-Participants

Variable / Scale	Median pre	Median post	z	Sig. (2-tailed)	Effect Size r
Organizing participants	52.08	58.33	-0.505	.613	-.06
Organizing non-participants	54.17	50.00	-2.576	.010**	-.16
Elaborating participants	54.17	45.83	-2.995	.003**	-.35
Elaborating non-participants	58.33	45.83	-6.713	.000***	-.43
Repeating participants	50.00	47.62	-2.627	.009**	-.31
Repeating non-participants	47.62	38.10	-5.298	.000***	-.32
Effort participants	66.67	62.50	-0.892	.373	-.10
Effort non-participants	66.67	62.50	-4.293	.000***	-.27
Attention participants	56.67	46.67	-1.213	.225	-.13
Attention non-participants	60.00	40.00	-3.639	.000***	-.21
Time Man. / Planning participants	41.67	54.17	-1.939	.052	-.23
Time Man. / Planning non-part.	41.67	43.75	-0.965	.334	-.06
Learning Env. participants	66.67	72.22	-0.113	.910	-.01
Learning Env. non-participants	61.11	66.67	-1.000	.317	-.06
Peer Learning participants	57.14	61.90	-0.466	.641	-.05
Peer Learning non-participants	57.14	59.52	-0.936	.349	-.06
Using Reference participants	83.33	75.00	-0.890	.374	-.09
Using Reference non-participants	75.00	66.67	-4.681	.000***	-.26

* $< .05$, ** $< .01$, *** $< .001$

However, looking at the mean scores[34], it becomes clear that MP^2-Math/Plus participants relate a higher level of attention than non-participants, both in the pre survey ($M_{pre,part.}$ = 51.59, $M_{pre,non-part.}$ = 58.04) and in the post survey ($M_{post,part.}$ = 42.86, $M_{post,non-part.}$ = 44.47), which means lower distraction respectively higher concentration for the project participants.

5.8.4 Summary

Only paired data sets from the pre and post surveys were entered into the exploration of the development of the use of learning strategies. *Attention, Elaborating, Repeating,* and *Using Reference* showed the most distinctive and significant changes, in the direction of less distraction (i.e. more attention), and, counter-intuitive, less frequent use of the other strategies. This tendency is less marked for MP^2-Math/Plus participants, except for *Repeating,* which shows no different development when participants are compared to non-participants. For *Time Management / Planning,* project participants report a development that opposes the general tendency: they show more frequent use of this strategy. In reference to *Effort,* MP^2-Math/Plus participants report no change, whereas the non-participants' scores decrease. Project participants show a generally higher level of *Attention,* though their concentration lessens, too.

---

[34] The statistics for the paired sample are attached in Appendix J.

6 Project Evaluation: Summary and Discussion

This chapter summarises and discusses the results from Chapter 5 and the findings from Chapter 4 in front of the background of the research approach and the research questions from Chapter 3, see orientation figure below.

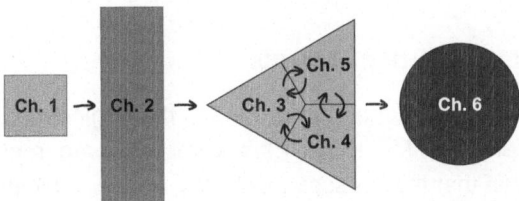

Section 6.1 discusses the conception and implementation of MP^2-Math/Plus, followed by section 6.2 which covers the impact of learning strategies. How these develop in the course of the first months at university is debated in section 6.3. Last, the research procedure employed is reflected critically in section 6.4, and the chapter closes with recommendations for further work.

6.1 Project Conceptualisation and Development

Our study object, the project MP^2-Math/Plus, is well-established at Ruhr-Universität Bochum. It comprises interventions to support first-year engineering students in mathematics addressing weaker students who have realised the necessity to adapt their learning behaviour to university standards. The first four cycles of MP^2-Math/Plus have seen some changes in the project procedures, which led to an overall concept that can be summarised, according to the *Design Research* framework by van den Akker (2013, cf. p. 67) as follows: *If you want to design a support intervention for first-year engineering students in mathematics, then you are well-advised to characterise the intervention by a focus on learning behaviour, and to do so via an interconnected collection of procedures that attend not only to subject-specific obstacles, but that also incorporate affective factors referring to effort and motivation.*

This section covers the conceptualisation of MP^2-Math/Plus, its design, and development, and thus answers the first three research questions.

RQ1 What procedures can specifically support first-year engineering students in mathematics?

RQ2 What are the characteristics of an intervention supporting first-year engineering students in mathematics?

RQ3 What combination of interventions is appropriate to promote learning strategies (as well as motivation)?

6.1.1 First-year engineering students

The needs and the educational background of the engineering first-years who form the target group of MP^2-Math/Plus were assessed in sections 5.1 and 5.2. We have seen that they possess a slightly weaker educational basis than the complete cohort, as they were less likely to have attended a *Gymnasium*[1] or an advanced course in mathematics at school, although the school marks show no distinctive differences. The students applying for MP^2-Math/Plus had visited the university's preparation course more often than average and there were more females than their percentage in the course would suggest. Apart from the gender distinction, these facts are in line with the intended target group of MP^2-Math/Plus, as the group of mediocre students in need of support (expressed in the weaker school background) who were aware of their deficiencies (manifested in their readiness to invest extra time to attend the preparation course) is aimed at.

The needs and wishes expressed by the applicants themselves focus on mathematical support and often mirror their desperation in the face of tertiary mathematics (see section 5.1). This corresponds to the stimulus in the transactional model of stress and coping by Lazarus (1991), see Figure 2.5, and also expresses emotions, which McLeod (1992) sees as central for

---

[1] Attendance of a *Gymnasium* as such does not imply a higher standard of mathematics education, as particularly in the last three years, students at a *Gesamtschule* are taught to the same standards. Nevertheless, the *Gymnasium* is the school usually chosen for pupils whose academic aptitude leaves no doubt at the age of 10.

the learning of mathematics, see section 2.1.2. The explicit wish to find help forms an ideal prerequisite for a support project which offers coping techniques and resources that can enable the individual to reappraise the situation and to adapt their learning behaviour. This interpretation supports the decision to let students experience university life for some time before the project start. When the stimulus originates in an official feedback in the form of an achievement test (like in the first project cycles), students' subjective assessment can build on that. But even without external feedback the individual may subjectively assess the situation as a threat or a challenge.

According to Lazarus (1991), it then matters to influence the second appraisal of the potentially stressful situation in a way that the resources are perceived as sufficient. This can work out if the learners judge their competence[2] as acceptable. Therefore, they should experience the beneficial impact of their actions (learning behaviour) on the outcome, the learning result, ideally understanding and the successful solving of tasks. According to Boekaerts (1999), (Brown, 1987) and others referred to in sections 2.1.2 and 2.1.2, this is not a purely rational process, but concerns affective aspects as well. Deci and Ryan (1990, 2000) put psychological needs in the centre: apart from competence, they list autonomy and relatedness. As many of the procedures of MP^2-Math/Plus are connected to social contacts, e.g. the *Learning Groups* (see section 4.5.1), the MP^2-Math/Plus *HelpDesk* (see section 4.5.10), or social networks (see sections 4.5.6 and 4.6.3), relatedness is well-established. Autonomy, the need to perceive oneself as actively making decisions, is compromised, though: Participants of MP^2-Math/Plus can adapt their learning behaviour as they choose, after they have tried out the suggestions propagated in the sessions of the *Learning Group*.

6.1.2 Selected project procedures

The fact that the majority of applicants express mathematical support as the main incentive for their applications (see section 5.1) is accommodated into the project procedures in various ways. The decision to try out learning strategies in the MP^2-Math/Plus *Learning Groups* by taking the example of mathematical contents from the current lecture has proved itself in practise, and combines

---

[2] In the sense of Deci and Ryan (1990, 2000), as the experience of control over what is considered relevant.

the needs of the participants with the central concept idea of the project. Occasionally there was even critique that the focus on learning strategies was too pronounced. In practise, it is a challenge to satisfy all wishes, but these utterances point at the fact that without a close connection to the mathematical subject matter, some participants would not have embraced new kinds of learning behaviour at all. The staff and effort put into the MP^2-Math/Plus-*HelpDesk* (section 4.5.10) are an important aspect for this, and they are appreciated by the participants accordingly (see section 5.3). Whenever a university establishes a help project for mathematics, however, a help desk or learning centre is usually among the interventions (see section 2.1.5); it is a common notion. The MP^2-Math/Plus *HelpDesk* has realised its mission with comparatively little resources (several hours of student assistants' work), very limited space (seminar room), and no expenses for learning material (apart from scrap paper). More resources are needed to establish mathematical support in digital form, which is also attempted in MP^2-Math/Plus in various ways (see sections 4.5.5 and 4.6.2). This could be even more elaborated and detailed and more apposite to specific user groups, however, and the options for feedback have not yet been exhausted.

The MP^2-Math/Plus *Revision Course* is another procedure that suits both the participants' need for mathematical support and the central characteristic of the project to address participants' learning behaviour and thus guide them towards responsibility for their learning. The last is true because of the timing of the course, after lectures have been completed and the bulk of repetition work should have been done, but well before the crucial examination. Had the *Revision Course* taken place during the semester, while lectures were still going on, it would not have contributed to students' abilities to come to grips with mathematics lectures, as students might have replaced the lectures with the *Revision Course*[3]. And if the *Revision course* had taken place during the very last days before the written examination, this could have tempted participants to sit back and wait for being passively fed the necessary skills without any activity on their part.

The *Mock Test* offered to the MP^2-Math/Plus participants was meant to enable them to live through the stress of an important examination, but without

---

[3] As a feedback to the *Revision Course*, some students actually said so, probably meaning it as a compliment for the down-to-earth explanations that helped them understand mathematics.

experiencing the negative consequences. Ideally, the *Mock Test* should have provided them with the certainty that they were well prepared. Realistically, a trial examination can never be truly authentic, though working without notes or formulary presents a challenge. And although the setting of the *Mock Test* was modelled on a real examination, it might not feel more authentic than the old examination papers that students had been practising with for quite some time. As such, the *Mock Test* is a dispensable project procedure, but as it implies no great expenditure, it augments the other procedures. Leaving aside its practical impact on students' learning behaviour, the MP^2-Math/Plus *Mock Test* has the potential to convey belonging to a group and encountering obstacles together – in the sense of Deci and Ryan, relatedness.

Two characteristics of MP^2-Math/Plus have evolved after the first pilot cycle: linking the project procedures and gradually withdrawing support (sections 4.5.1 and 4.5.2). Both are based on the perception that students, although adults, fare better with a certain level of obligation on the way to personal commitment and self-regulation. This is in keeping with the deliberations by Boekaerts (1999) that students' commitment is essential (section 2.1.2), and therefore needs to be advanced in case it is not fully developed. The issue in mathematics for engineering students is that for them, the self-chosen goal of becoming an engineer is not directly connected to the learning of mathematics, so further help is needed to make the connection[4]. In the self-determination continuum by (Deci & Ryan, 2000), see Figure 2.7, the motivation aimed at in MP^2-Math/Plus is located on the right (intrinsic motivation), but provisions are made against forms of extrinsic motivation placed in the middle with "somewhat external" (p. 237) forms of regulation. The goal is to help students advance towards more internal forms of self-regulation, so that they will proceed successfully through their education without further external regulation. As Lazarus (1991) mentions (see section 2.1.2), having successfully mastered stressful situations and experienced the desired outcome as a result of one's own choices and actions (thus attributing success internally, see Table 2.6) can favour future successful coping of similar challenges.

---

[4] Incidentally, this is also covered by MP^2-Math/Practice, where engineering students learn about the necessity of mathematics for practical engineering problems, see Härterich et al. (2012).

6.1.3 Students' progression at university

According to Lazarus (1991), participating in MP^2-Math/Plus may have helped students to master other modules of their engineering course. To explore if this is the case, if and how MP^2-Math/Plus participants are successful in their further studies, is the focus of section 5.4. The view on students' progression through their university course provides a new perspective of evaluating MP^2-Math/Plus. One of the aspects behind the project's focus on learning strategies is that these are universally applicable; the study routines and techniques compulsory for mathematics can prove conducive for other subjects, too. Planning your time, segmenting and structuring the subject material, integrating rewards and communication with others, or simply keeping your work environment and papers in order are not mathematics-specific; they come in useful everywhere. Consequently, data of students' examination records over the course of some years are of interest. The downside of this kind of long-term study is that the data is hard to come by – be it for reasons of data protection, or because it can never be complete[5]. So this data has to be interpreted with care. With regard to retention rates, Table 5.15 can be cautiously boiled down to the fact that, a few years after enrolment, there are less students from the MP^2-Math/Plus groups still matriculated, compared to the complete group. This confirms them as an at-risk group, even as a less successful group. The impression is supported by the fact that a smaller share of MP^2-Math/Plus participants masters all three mathematics and the two mechanics modules (24% for SDG respectively 25% for SLG) when compared to the complete cohort of *Mechanical Engineering* students (31%)[6], but the small absolute numbers forbid reliable conclusions. In actual fact, the interpretation that MP^2-Math/Plus project participants' progress through the course does not vary considerably from the complete group is more likely, see Figure 5.3 – which is an asset, taking into account that they refer to at-risk students in a higher share. We can tentatively conclude that MP^2-Math/Plus participants show no distinctive

---

[5] If a student is shown as "exmatriculated" in the files at RUB, he / she could have given up academic aspirations altogether, could have changed the subject, or could be successfully studying engineering at another university.

[6] These numbers are even lower than those reported by (Heublein et al., 2014), which is consistent with the fact that they refer to only one university, and do not take into account that someone who does not pass a course module there may still do so at another university or even change to a different course.

differences in their further progress through the engineering course modules, when compared to the complete cohort. Moreover, though the data base is unstable, it seems that the obstacles from the first semester (here *Mathematics 1* and *Mechanics A*) are the most difficult to overcome[7]. Once a student has passed these, the ensuing attrition rates are reasonable for all groups. This finding supports the MP^2-Math/Plus hypothesis that the first semester is crucial, and that it is advisable to schedule support projects then.

6.2 Impact of Learning Strategies

Research questions four and five refer to the relevance and the impact of MP^2-Math/Plus and are discussed in this and the following section.

RQ4 Which learning behaviour is connected with academic succss?

RQ5 How does MP^2-Math/Plus influence learning strategies (and motivation)?

6.2.1 Preliminary considerations

This section is dedicated to the influence of the different kinds of learning strategies on examination success. One of the basic assumptions of MP^2-Math/Plus is the hypothesis that learning strategies can influence learning outcome considerably. This was explored with linear models containing different learning strategies as predictors and academic success as outcome variable. As section 5.7 has revealed, they can explain up to 28% of variance, which can make a huge difference on the learning outcome. This needs to be viewed in connection to the fact that the LIST scales on learning strategies do not cover intellect or understanding. We have seen that, when the variable *Mini Test*, which describes mathematical competence at the beginning of the university course, is entered into the model, the explained variance rises to over 40%. Even on its own, *Mini Test* has a highly significant ($p = .000$) and very strong (standardised coefficient $\beta = -.634$, which implies $R^2 = .40$) influence on the result of the written examination months later. The negative β value means that the direction of the influence is as expected: The better the result in the mini exam (measured in percentage of points achievable), the better the mark in the

---

[7] It would not be adequate to say that *Mathematics 1* is more difficult than *Mechanics A* from these data, because the attrition rates are calculated in reference to the previous module.

written examination (measured in German school marks). Other researchers, too, have found that the knowledge and skills students have acquired at school is a strong predictor for their success at university (Rach, 2014; Schiefele, Streblow, Ermgassen, & Moschner, 2003; Trapmann et al., 2007). Because the focus of our research does not lie on variables that cannot be influenced, but rather on skills that can be improved by project procedures, this only serves as background to understand why learning strategies explain comparatively little variance. As MP^2-Math/Plus aims at supporting mediocre or weak students, a little help may just be the assistance these students need in order to pass their examination in mathematics. In this regard, the pass rates observed in section 5.5 are adequate, and in some instances gratifying.

Another aspect is relevant in this respect: Our survey (respectively the LIST questionnaire) asks how often a learning behaviour is employed, and not how efficiently it is applied. For example, the item *Whenever I did not understand a technical term, I looked it up in a textbook or on the Internet*[8] does not ask for effective use of the Internet (or other sources) for this purpose. A student might mark this item with *very often* when he / she laboriously searches the Internet for the definition of a term that may appear in different areas of mathematics, without narrowing the search down, thus calling up a growing number of websites without finding the desired definition[9]. And although e.g. *Elaborating* describes learning behaviour where the new subject matter is attempted to be integrated into a network of already known material, the existence of a viable previous network is uncertain. *Elaborating* and *Organizing* describe the learning strategies that would primarily be categorised as covering intellectual comprehension and understanding. Surprisingly, they do not correlate with *Academic Success* in a relevant or even significant way, see Table 5.34. Consequently, they do not show up in any of the linear models.

6.2.2 Interpretation of the final model

The final model (see Table 5.38) contains only the two predictors *Effort* ($\beta = -.162$) and *Attention* ($\beta = .149$). The importance of *Effort* is additionally

---

[8] Item 12 (b) from the *Reference* scale, see Appendix B.

[9] For example, looking up *dot product* in English Wikipedia alone will render information on triple product extension, dyadics, and tensors – none of which are relevant for first-year engineering mathematics; the German version of the Wikipedia page on *Skalarprodukt* is linked to L^2- and Hilbert spaces.

stressed by the fact that it is the only remaining scale when restricting the model to MP^2-Math/Plus project participants, see Table 5.39. This is reason enough to revisit the items that constitute the *Effort* and *Attention* scales, see Appendix B or Appendix I. All these items concern making an effort, not giving up, addressing difficult tasks, or taking pains and investing time to study. As presented in section 2.2.2, the *Effort* and *Attention* scales unite items from other questionnaires (particularly MSLQ) that specifically relate to motivation there, though LIST itself does not have a scale called *Motivation*. These findings back the decision described in section 4.5, that apart from the technical side of learning behaviour, the project procedures should be characterised by a focus on commitment. Besides, the exceptional position of *Effort* supports the analysis from section 2.1.4 that mathematics lectures for engineering students concentrate less on abstract concepts and proofs, and more on calculations and computations, which are within reach for a high-school graduate, regardless of mathematical talent. It also contradicts the appraisal that "learning mathematics is a question more of ability than of effort" (McLeod, 1992, p. 575), see section 2.1.2. The items from the *Attention* scale (see Appendix B or Appendix I) are all reverse-coded (hence the positive β values) and refer to losing one's concentration, being distracted, or thoughts straying. That places them in a similar context as the items from the *Effort* scale, but the investigations in section 5.6 have shown that they load on a separate factor, and section 5.7 has delivered the finding that *Effort* and *Attention* as predictors in a linear model do not implicate a collinearity problem. Consequently, *Attention* specifies learning behaviour different from *Effort*. And indeed, whereas *Effort* is more about the intention to work hard, *Attention* covers the execution of these ideas. In this respect, the fact that the *Attention* scale loses its importance when the linear model is restricted to MP^2-Math/Plus participants (see Table 5.39), can be interpreted as project participants' eschewal of inattention. In other words: If for project participants, the level of *Attention* is homogeneously similar and it is not a relevant factor for their academic success, they do not have an attention problem.

The conclusion that *Effort* (or motivation), together with *Attention* is *the* central factor in mathematics for engineering students is also a reason for optimism, as this is something that can be addressed by project interventions. On the one hand, it is important to recruit students who bring with them the requirement of being motivated for their course. On the other hand, this motivation needs upkeep and cultivation. This is exactly where a project like MP²-Math/Plus can act out its strengths, e.g. in creating relatedness, autonomy, and competence, in the diction of Deci and Ryan (1990), as discussed in section 6.1, and a strong factor in recommending this kind of project work.

The importance of *Effort* is also in keeping with other research: Schiefele et al. (2003) found that effort had the strongest influence on academic success (R^2 = .26) in a structural equation model containing items on motivation, on self-concept, on the (perceived) quality of teaching, on epistomological beliefs, and items on learning strategies, and that effort "mediated the effect of all other variables on achievement" (p. 185). What is more, they found that effort was determined by achievement motivation[10] with a standardised coefficient of .43 and by interest in studying[11] with .15.

In spite of its relevance, our data does not yield the expected mediator role of *Effort*, though (see section 5.7.5). This may have its reasons in the fact that apart from the learning behaviour covered by the LIST questionnaire, other factors may play a role, or the relationship may not be simply linear, which is in keeping with Eley and Meyer (2004) who found that the use of different learning strategies, such as "the relating of mathematics to real world contexts, the systematic and principled use of examples, the adoption of overall planning and goal setting approaches, the meta-level reflection on procedures, and in contrary direction, the use of mindless substitution, rule application, and repetition" (p. 449) related to mathematics performance, but "the patterns of these relationships were not simple monotonic linear" (p. 449).

There is evidence that the importance of effort can vanish when other factors are introduced. Our own research (Griese & Kallweit, 2016) found that continuous effort had no significant influence on academic success (standardised $\beta = -.09$, $p = .36$) when integrated in a model with weekly assignments, lecture attendance and subsequent work, tutorials, surface learning, and deep learning techniques. There, the weekly assignments presented the strongest influence with $\beta = .51$, which was highly significant.

---

[10] German: *Leistungsmotivation*
[11] German: *Studieninteresse*

6.3 Development of Learning Strategies

The next step after establishing which learning behaviour implies successful knowledge acquisition is to explore which learning behaviour changes or correlates with project participation. For MP^2-Math/Plus, this was addressed by collecting data from engineering first-years at the beginning and at the end of their first semester, matching it to examination results and project participation, and comparing the scores, see section 5.8.

6.3.1 Reflections on the pre-post scenario

Employing the LIST questionnaire in pre-post surveys involved more data sets in the pre surveys than in the post surveys, see Table 5.28. This is connected to the fact that not all students present during the survey at the beginning of the semester were also present at the end – and vice versa. A bias towards desired learning behaviour in post surveys is possible, as those students still attending lectures at the end of the semester might also show well-adjusted conduct in other areas. But it is also feasible to observe a bias towards more eager learning performance in pre surveys, as at the beginning of a course, students might tend to still be enthused with their new situation on life and might report their good resolutions. It is also thinkable that the self-reported frequency of how often a learning behaviour was employed suffered a change of reference frame, and what was considered as *often* while having school demands in mind is rated *seldom* after some months at university. Which (if any) interpretation is appropriate cannot be concluded from the data.

The decision to analyse only paired data sets ensured a certain validity, although their scores differ slightly from the unpaired data. The fact that the paired data yields higher pre scores is not surprising, as the paired data stems from students who persevered until the end of the semester, so their more ambitious learning behaviour is consistent. Selecting only data sets that can be paired seems to have eliminated students with less frequent use of learning strategies. This does not contraindicate the planned explorations, though.

6.3.2 Learning strategies related to motivation

The items from the LIST *Effort* and *Attention* scales reflect the motivation items from MSLQ[12] (section 2.2). So, in order to explore how MP²-Math/ Plus influences motivation, the *Effort* and *Attention* scales and their development from the beginning to the end of the first semester (from the pre to the post survey) provide information, see section 5.8[13]. We have seen there that the scores for these scales are lower in the post survey than in the pre survey, which means different things: For *Effort*, a lower score means less effort is made – or at the very least, that the students subjectively assess that they are making less effort, in comparison to an unspecified ideal. This ideal itself may have shifted from the beginning of their studies until a few months later, when they were realising that university makes higher demands than anticipated. This description is true both for the complete cohort and for the participants of MP²-Math/Plus. To be exact, the change in *Effort* is only insignificantly smaller for project participants (mean difference = 2.46, SE = 2.35, $t(38)$ = 1.046, p = .302), whereas the complete group undergoes a significant decrease (mean difference = 5.78, SE = 1.24, $t(123)$ = 4.674, p = .000***, r = .39) with medium effect, see Table 5.47. This means that in comparison, MP²-Math/Plus participants keep their level of motivation, whereas the average student's motivation decreases.

For *Attention*, however, the situation presents itself as follows: As the items of this scale are reverse-coded, a higher score means more distraction or less concentration. The fact that the *Attention* scores are lower in the post survey means that students report a higher level of concentration at the end of the semester. The change of *Attention* is more distinct among the non-participants, but the MP²-Math/Plus participants start at a higher level of concentration and end at roughly the same level as the non-participants, see Table 5.47 and Figure J.3. So we can summarise that for both scales connected with motivation, MP²-Math/Plus participants show advantages over the non-participants, which can be understood as a reason for their similar advance through the mathematics and mechanics modules, in spite of their weaker educational background.

---

[12] Motivated Strategies for Learning Questionnaire, (Pintrich et al., 1993), see section 2.2.1.
[13] Our data implied some difficulties with model fit (see section 5.6), but these are restricted to the metacognitive scales and do not pose a problem here.

6.3.3 Other learning strategies

After *Effort* and *Attention*, the other LIST scales are now discussed in declining order of markedness (see Table 5.42). At the end of the semester, students reported significantly less use of the learning strategy *Elaborating* ($M = 46.27$, $SD = 18.83$) than at the beginning of the semester ($M = 57.71$, $SD = 17.04$, $t(160) = 8.773$, $p = .000***$, $r = .57$), see Table 5.45. This can be understood before the background that *Elaborating* covers learning strategies that are more expedient when reviewing and summarising the learning material, typically during the semester. At the time of the post survey, these activities may have been already completed, or thrust aside in favour of surface learning strategies. There is no significant difference for *Elaborating* between MP^2-Math/Plus participants and non-participants (see Table 5.47).

Repeating shows decrease of use from the beginning until the end of the semester, too, but there is a difference between MP^2-Math/Plus participants and non-participants (from median$_{pre}$ = 50.00 to median$_{post}$ = 47.62, $z = -2.627**$, $r = -.31$ for the participants, and from median$_{pre}$ = 47.62 to median$_{post}$ = 38.10, $z = -5.298***$, $r = -.32$ for the non-participants), see Table 5.48: The use of *Repeating* strategies decreases less among the participants. Although *Repeating* covers surface learning strategies such as memorising formulae or rereading notes, their usefulness shortly before an examination is undoubted. The fact that MP^2-Math/Plus participants have learned about memorising techniques in the project might have made a contribution here, too.

Concerning *Using Reference*, the conclusion is similar: The decrease in the use of this learning strategy is smaller among the participants (from median$_{pre}$ = 83.33 to median$_{post}$ = 75.00, $z = -0.890$, $p = .374$ for the participants, and from median$_{pre}$ = 75.00 to median$_{post}$ = 66.67, $z = -4.681***$, $r = -.26$ for the non-participants), see Table 5.48. For these two scales, MP^2-Math/Plus participants report a higher frequency of use of these learning strategies at the beginning of the semester, which decreases, but only to the level of the non-participants in the pre survey (in reference to the medians). As remarked in section 6.2, an extensive use of reference works or the Internet does not necessarily imply a more efficient practical effect, nonetheless MP^2-Math/Plus participants have learned and practised search techniques that swiftly lead to the desired information. The remaining scales do not yield any distinctive changes.

6.3.4 Résumé

We have seen that participation in MP^2-Math/Plus has a positive effect on learning behaviour connected with motivation. And although not all of the scales resulting in favourable interpretations for MP^2-Math/Plus have an attested positive impact on the outcome of the mathematics examination (see section 6.2), these findings can still be regarded as an asset, because they show that participation in the project has made a difference. If this difference had been as pronounced without MP^2-Math/Plus' procedures is doubtful, but we cannot be certain as there is no viable data for a control group. The initially planned *Monitored Group* vanished due to reasons of project design and enhanced resources, see section 4.6. If it had been kept, though, the consequences of being rejected support and being denied partaking in a *Revision Course* and visiting the *HelpDesk* could have influenced students' motivation negatively. Now that the relevance of motivation has become clear for our target group, we can conclude that this would probably have enhanced rather than diminished the differences to the project group, resulting in the opposite effect a control group should have.

6.4 Concluding Remarks and Outlook

We have seen that a combination of interconnected procedures covers different aspects of the support required by engineering first-years in reference to subject-specific help and to affective aspects. The results are satisfactory with respect to the development of learning strategies (particularly LIST scales *Repeating* and *Using Reference*) and motivation (*Effort* and *Attention*), and subsequently to their examination results. Concerning the progress through further course modules, the numbers are acceptable. In reference to Plomp and Nieveen (2013b), MP^2-Math/Plus presents itself as a "complete and final" intervention (p. VIII) after adaptations of the first project cycles. It fulfils the requirements of being "relevant", "consistent", "practical", and "effective" (p. VIII), see section 3.2. The importance of non-cognitive aspects such as motivation, effort, and attention has been highlighted. Compared to them, the activities usually connected with the learning and understanding of mathematics (e.g. finding examples and connections, visualising, compiling summaries) have lost their significance. This is presumably specific for engineering mathematics. When expanding the project, a precise fit for the intended target groups is imperative[14].

The decision for *Design Research* as the theoretical scaffold for our research can now be reviewed. It has proved well-suited for an educational project spanning several cycles, as it allows for adaptations from one cycle to the other. In slight deviance to the model, in practise the major adaptations were necessary after the first cycle; later only minor changes were made. The demand to consider the purpose, the context, the characteristics, and the procedures separately, as postulated by van den Akker (2013), helped to structure the conceptualisation and reflection processes. The fact that there are well-researched examples of successful *Design Research* projects available (e.g. Plomp & Nieveen, 2013b) is a further asset. That MP^2-Math/Plus does not significantly contribute to the development of new theories (although our results confirm the importance of affective factors in the learning of mathematics) can be regarded as a drawback in the decision for *Design Research*, as the requirement of a balance between theory and practise is not fully met.

Our methodological procedures proved suitable for the purpose. The review of questionnaires on learning strategies left little alternative to the German LIST questionnaire, although from the gathered results, now a more differentiated instrument for measuring motivation seems advisable when researching service mathematics. Measuring tools available include questionnaires like in Rheinberg, Vollmeyer, and Burns (2001), but other forms like a Thematic Apperception Test (Morgan, 2002; Murray, 1943) might also be suitable. Other target groups might profit from a questionnaire also covering more mathematics-specific learning strategies, like attempted in Liebendörfer et al. (2014). The pre-post surveys were deemed appropriate for describing development, but the unexpected general decrease in the use of learning strategies posed difficulties. Some authors (e.g. Nimon, Zigarni, & Allen, 2011; Lam & Bengo, 2003) thus favour using a retrospective test design to measure programme effectiveness, which can be just as adequate (as the results in Griese & Kallweit, 2016, show)

---

[14] For example, MP^2-Math/Plus' sister project for B.A. students has accommodated different procedures, mostly in the direction of a clearer focus on the subject matter, on visualisation, and on catching up on basic algebra skills from middle school.

and reduce data collection and matching problems. Analysing qualitative data from interviews or learning diaries could complement our findings and therefore are a recommendation for future research approaches.

On the whole, MP^2-Math/Plus has demonstrated its worth. Apart from the active support of students willing to apply themselves, the ensuing exchange among lecturers, teaching assistants, and researchers can be a basis for the improvement of teaching, both via an awareness for affective factors and through an emphasis on learning strategies. Projects covering a similar range of aspects are highly recommended. Refining the project procedures, particularly for other target groups, can consequently be a focus of further work. Longitudinal studies on the development of engineering mathematics with respect to the SEFI curriculum (Alpers, 2014, 2016) add another aspect, as more competence-orientation might change the efficiency of certain learning strategies. The expansion and consolidation of digital support tools, particularly in reference to goal-setting and feedback ideas, seem also promising. Digitalisation would additionally facilitate collaboration with and transfer to other universities who face similar challenges.

Our findings thus lead the way to further research and collaboration in this field. Its many facets hold the prospect of more worthwhile results.

References

Aebli, H. (1976). *Grundformen des Lehrens: Eine allgemeine Didaktik auf kognitionspsychologischer Grundlage* (9th ed.). Stuttgart: Klett.

Alpers, B. (2014). *A mathematics curriculum for a practice-oriented study course in mechanical engineering, version 1.3.* Aalen. Retrieved 23.03.2016, from http://sefi.htw-aalen.de/Curriculum/Mathematics_curriculum_for_mechanical_engineering_February_3_2014.pdf

Alpers, B. (2016). Das SEFI Maths Working Group "Curriculum Framework Document" und seine Realisierung in einem Mathematik-Curriculum für einen praxisorientierten Maschinenbaustudiengang. In A. Hoppenbrock, R. Biehler, R. Hochmuth, & H.-G. Rück (Eds.), *Lehren und Lernen von Mathematik in der Studieneingangsphase* (pp. 645–659). Wiesbaden: Springer Spektrum.

Alpers, B., et al. (Eds.). (2013). *A framework for mathematics curricula in engineering education.* Brussels: SEFI.

Alrø, H., Ravn, O., Valero, P., & Skovsmose, O. (2010). *Critical mathematics education: Past, present and future: Festschrift for Ole Skovsmose.* Rotterdam and Boston: Sense.

Anderson, T., & Shattuck, J. (2012). Design-based research: A decade of progress in education research? *Educational Researcher, 41*(1), 16–25.

Arnon, I., Cottrill, J., Dubinsky, E., Oktac, A., Roas Fuentes, S., Trigueros, M., & Weller, K. (2014). *APOS Theory: A framework for research and curriculum development in mathematics education.* New York, Heidelberg, Dordrecht, London: Springer.

Artigue, M., Batanero, C., & Kent, P. (2007). Mathematics thinking and learning at post-secondary level. In F. K. Lester (Ed.), *The handbook of research on mathematics education* (pp. 1011–1049). Greenwich, Connecticut: Information Age Publishing.

Artino, A. R. (2005). *A review of the Motivated Strategies for Learning Questionnaire.* ERIC documents ED499083. Retrieved 12.01.2014, from http://eric.ed.gov/?id=ED499083

Attorps, I., Björk, K., Radic, M., & Viirmann, O. (2013). Teaching inverse functions at tertiary level. In B. Ubuz, C. Haser, & M. A. Mariotti (Eds.), *Proceedings of the Eighth Conference of the European Society for Research in Mathematics Education* (pp. 2524–2533). Ankara: Middle East Technical University.

Bandura, A. (1977). *Social learning theory.* Englewood Cliffs and N.J: Prentice Hall.

Barab, S., & Squire, K. (2004). Design-based research: Putting a stake in the ground. *Journal of the Learning Sciences, 13*(1), 1–14.

Bargel, T. (2015). *Studieneingangsphase und heterogene Studentenschaft - neue Angebote und ihr Nutzen: Befunde des 12. Studierendensurveys an Universitäten und Fachhochschulen* (Vol. 83). Konstanz: Arbeitsgruppe Hochschulforschung.

Bathke, G.-W., Schreiber, J., & Sommer, D. (2000). *Soziale Herkunft deutscher Studienanfänger: Entwicklungstrends der 90er Jahre.* Hannover: HIS.

Baumert, J. (2001). *PISA 2000: Basiskompetenzen von Schülerinnen und Schülern im internationalen Vergleich.* Opladen: Leske + Budrich.

Becker, R., Haunberger, S., & Schubert, F. (2010). Studienfachwahl als Spezialfall der Ausbildungsentscheidung und Berufswahl. *Zeitschrift für ArbeitsmarktForschung, 42*(4), 292–310.

Bempechat, J., Nakkula, M. J., Wu, J. T., & Ginsburg, H. P. (1996). Attributions as predictors of mathematics achievement. *Journal of Research and Development in Education, 29*(2), 53–59.

Berger, M., & Schwenk, A. (2006). Zwischen Wunsch und Wirklichkeit: Was können unsere Studienanfänger? *Die Neue Hochschule*(2), 36–40.

Bikner-Ahsbahs, A., Dreyfus, T., Kidron, I., Arzarello, F., Radford, L., Artigue, M., & Sabena, C. (2010). Networking of theories in mathematics education. In M. M. F. Pinto & T. F. Kawasaki (Eds.), *Proceedings of the 34th Conference of the International Group for the Psychology of Mathematics Education* (pp. 145–175). Belo Horizonte, Brazil: PME.

Bildungsministerium für Bildung und Forschung (Ed.). (2007). *Zur Entwicklung nationaler Bildungsstandards.* Bonn, Berlin: BMBF. Retrieved 27.10.2015, from www.bmbf.de/pub/zur_entwicklung_nationaler _bildungsstandards.pdf

Blömeke, S. (2016). Der Übergang von der Schule in die Hochschule: Empirische Erkenntnisse zu mathematikbezogenen Studiengängen. In A. Hoppenbrock, R. Biehler, R. Hochmuth, & H.-G. Rück (Eds.), *Lehren und Lernen von Mathematik in der Studieneingangsphase* (pp. 3–13). Wiesbaden: Springer Spektrum.

Blömeke, S., Gustafsson, J.-E., & Shavelson, R. J. (2015). Beyond dichotomies: Competence viewed as a continuum. *Zeitschrift für Psychologie, 223*(1), 3–13.

Blum, W., Drüke-Noe, C., Hartung, R., & Köller, O. (2006). *Bildungsstandards Mathematik: konkret.* Berlin: Cornelsen Scriptor.

Boekaerts, M. (1997). Self-regulated learning: A new concept embraced by researchers, policy makers, educators, teachers, and students. *Learning and Instruction, Vol. 7*(2), 161–186.

Boekaerts, M. (1999). Self-regulated learning: Where we are today. *International Journal of Educational Research, 31*, 445–457.

Boekaerts, M., Pintrich, P. R., & Zeidner, M. (Eds.). (2000). *Handbook of self-regulation.* San Diego, London: Academic Press.

Breen, R., & Goldthorpe, J. H. (1997). Explaining educational differentials: Towards a formal rational action theory. *Rationality and Society*(9), 275–305.

Brown, A. L. (1987). Metacognition, executive control, self-regulation, other more mysterious mechanisms. In F. E. Weinert & R. Kluwe (Eds.), *Metacognition, motivation, and understanding* (pp. 65–109). Hillsdale, NJ: Erlbaum.

Brown, A. L. (1992). Design experiments: Theoretical and methodological challenges in creating complex interventions in classroom settings. *The Journal of the Learning Sciences, 2*(2), 141–178.

Bruder, R., Elschenbroich, J., Greefrath, G., Henn, H.-W., Kramer, J., & Pinkernell, G. (2010). Schnittstelle Schule - Hochschule. *Beiträge zum Mathematikunterricht*, 75–82.

Buchsteiner, J., & Kallweit, M. (2015). Professionalisierung des Helpdesk Mathematik. In F. Caluori, H. Linneweber-Lammerskitten, & C. Streit (Eds.), *Beiträge zum Mathematikunterricht 2015.* Münster: WTM.

Bühner, M. (2011). *Einführung in die Test- und Fragebogenkonstruktion* (3rd ed.). München: Pearson Studium.

Bundesministerium für Bildung und Forschung. (n.d.). *Daten-Portal.* Retrieved 12.10.2015, from http://www.datenportal.bmbf.de/portal/de/K253.html

Burkhardt, H. (2006). From design research to large-scale impact: Engineering research in education. In J. van den Akker, K. Gravemeijer, S. McKenney, & N. Nieveen (Eds.), *Educational design research* (pp. 121–150). London, New York: Routledge.

Cano, F. (2006). An in-depth analysis of the Learning and Study Strategies Inventory (LASSI). *Educational and Psychological Measurement, 66*(6), 1023–1038.

CERME 7. (2011). *Papers of Working Group 14, University Mathematics Education.* Retrieved 12.10.2015, from http://www.cerme7.univ.rzeszow.pl/index.php?id=wg14

CERME 8. (2013). *Papers of Working Group 14, University Mathematics Education.* Retrieved 12.10.2015, from http://cerme8.metu.edu.tr/wgpapers/wg14_papers.html

CERME 9. (2015). *Papers of Working Group 14, University Mathematics Education.* Retrieved 12.10.2015, from http://www.cerme9.org/products/twg14-under-construction/

Cervone, D., Mor, N., Orom, H., Shadel, W. G., & Scott, W. D. (2011). Self-efficacy beliefs and the architecture of personality: On knowledge, appraisal, and self-regulation. In K. D. Vohs & R. F. Baumeister (Eds.), *Handbook of self-regulation* (pp. 461–484). New York: Guilford Press.

Clark, M., & Lovric, M. (2008). Suggestion for a theoretical model for secondary-tertiary transition in mathematics. *Mathematics Education Research Journal, 20*(2), 25–37.

Clark, M., & Lovric, M. (2009). Understanding secondary-tertiary transition in mathematics. *International Journal of Mathematics Education in Science and Technology, 40*(6), 755–776.

Clements, D. H., & Sarama, J. (2007). Early childhood mathematics learning. In F. K. Lester (Ed.), *Second handbook of research on mathematics teaching and learning* (pp. 461–555). New York: Information Age Publishing.

Colberg, C., Biehler, R., Hochmuth, R., Schaper, N., Liebendörfer, M., & Schürmann, M. (08.03.2016). *Wirkung und Gelingensbedingungen von Unterstützungsmaßnahmen für mathematikbezogenes Lernen in der Studieneingangsphase.* Heidelberg: Jahrestagung der GDM.

Collins, A. (1992). Toward a design science of education. In E. Scanlon & T. O'Shea (Eds.), *New directions in educatinal technology* (pp. 15–22). New York: Springer.

Danckwerts, R., & Vogel, D. (2006). *Analysis verständlich unterrichten.* Berlin, Heidelberg: Spektrum Akademischer Verlag.

David, A. (2013). *Aufgabenspezifische Messung metakognitiver Aktivitäten im Rahmen von Lernaufgaben* (Dissertation). Technische Universität, Chemnitz.

Davydov, V. V. (1990). *Types of generalisation in instruction: Logical and psychological problems in the structuring of school curricula.* Reston, VA: National Council of Teachers of Mathematics. [Originally published in 1972].

de Guzmán, M., Hodgson, B. R., Robert, A., & Villani, V. (1998). Difficulties in the passage from secondary to tertiary education. In G. Fischer & U. Rehmann (Eds.), *Proceedings of the International Congress of Mathematicians* (pp. 747–762). Rosenheim: Geronimo.

DeBellis, V. A., & Goldin, G. (1997). The affective domain in mathematical problem solving. In E. Pehkonen (Ed.), *Proceedings of the 21st Conference of the International Group for the Psychology in Mathematics Education, Vol. 2* (Vol. 2, pp. 209–216). Lahti, Finland: PME.

Deci, E. L., & Ryan, R. M. (1990). *Intrinsic motivation and self-determination in human behavior* (3rd ed.). New York: Plenum Press.

Deci, E. L., & Ryan, R. M. (2000). The What and Why of goal pursuits: Human needs and the self-determination of behavior. *Psychological Inquiry, 11*(4), 227–268.

Dehling, H., Glasmachers, E., & Härterich, J. (2012). Mathematik im Doppelpack. *duz-Akademie, 4,* 5.

Dehling, H., Glasmachers, E., Härterich, J., & Hellermann, K. (2010). MP^2 - Mathe/Plus/Praxis: Neue Ideen für die Servicelehre. *Mitteilungen der Deutschen Mathematiker-Vereinigung, 18,* 252.

Design-Based Collective. (2003). Design-based research: An emerging paradigm for educational enquiry. *Educational Researcher, 32*(1), 5–8.

Dreyfus, T. (1991). Advanced mathematical thinking processes. In D. O. Tall (Ed.), *Advanced mathematical thinking* (Vol. 11, pp. 25–41). Dordrecht and Boston: Kluwer Academic Publishers.

Dreyfus, T. (Ed.). (1995). *Advanced mathematical thinking [Special Issue]* (Vol. 29). Dordrecht: Kluwer Academic Publishers.

Dreyfus, T. (2012). *Constructing abstract mathematical knowledge in context.* Retrieved from http://www.icme12.org/upload/submission/1953_F.pdf

Dubinsky, E., & McDonald, M. A. (2001). APOS theory: A constructivist theory of learning in undergraduate mathematics education research. In D. A. Holton (Ed.), *The teaching and learning of mathematics at university level* (pp. 275–282). Dordrecht and Netherlands: Kluwer Academic Publishers.

Duncan, T. G., & McKeachie, W. J. (2005). The making of the Motivated Strategies for Learning Questionnaire. *Educational Psychologist, 40,* 117–128.

Dunn, K. E., Lo, W.-J., Mulvenon, S. W., & Sutcliffe, R. (2012). Revisiting the motivated strategies for learning questionnaire: A theoretical and statistical reevaluation of the metacognitive self-regulation and effort regulation subscales. *Educational and Psychological Measurement, 72,* 312–331.

Edelson, D. C. (2006). Balancing innovation and risk: Assessing design research proposals. In J. van den Akker, K. Gravemeijer, S. McKenney, & N. Nieveen (Eds.), *Educational design research* (pp. 100–106). London, New York: Routledge.

Eisenberg, T. (1991). Functions and associated learning difficulties. In D. O. Tall (Ed.), *Advanced mathematical thinking* (Vol. 11, pp. 140–152). Dordrecht and Boston: Kluwer Academic Publishers.

Eley, M. G., & Meyer, J. H. F. (2004). Modelling the influences on learning outcomes of study processes in university mathematics. *Higher Education, 47*(4), 437–454.

Ellis, R. A., Goodyear, P., Rafael, A. C., & Prosser, M. (2008). Engineering students' conceptions of and approaches to learning through discussions in face-to-face and online contexts. *Learning and Instruction, 18,* 267–282.

Engelbrecht, J. (2010). Adding structure to the transition process to advanced mathematical activity. *International Journal of Mathematics Education in Science and Technology, 41*(2), 143–154.

Entwistle, A., & Entwistle, N. (1992). Experience of understanding in revising for degree examinations. *Learning and Instruction*(2), 1–22.

Entwistle, N., & McCune, V. (2004). The conceptual bases of study strategies inventories. *Educational Psychology Review, 16*(4), 325–346.

Entwistle, N., & Ramsden, P. (1983). *Understanding student learning.* London: Croom Helm.

Erdem Keklik, E., & Keklik, I. (2013). Motivation and learning strategies as predictors of high school students' math achievement. *Cukurova University Faculty of Education Journal, 42*(1), 96–109.

Europäische Kommission. (2009). *ECTS Leitfaden.* Brüssel: GD Bildung und Kultur.

Evans, J. S. B. T. (2007). *Hypothetical thinking: Dual processes in reasoning and judgement.* Hove: Psychology Press.

Fennema, E., & Leder, G. (1990). *Mathematics and gender.* New York: Teachers College, Columbia University.

Field, A. (2009). *Discovering statistics using SPSS* (3rd ed.). Thousand Oaks, CA: Sage Publications.

Fischbein, E. (1987). *Intuition in science and mathematics: An Educational Approach.* Dordrecht: Kluwer Academic Publishers.

Fischer, A., Heinze, A., & Wagner, D. (2009). Mathematiklernen in der Schule - Mathematiklernen an der Hochschule: Die Schwierigkeiten von Lernenden beim Übergang ins Studium. In A. Heinze & M. Grüßing (Eds.), *Mathematiklernen vom Kindergarten bis zum Studium* (pp. 245–264). Münster: Waxmann.

Forgasz, H. J. (1995). Gender and the relationship between affective beliefs and perceptions of grade 7 mathematics classroom learning environments. *Educational Studies in Mathematics, 28*, 219–239.

Fox, E., & Riconscente, M. (2008). Metacognition and self-regulation in James, Piaget, and Vygotsky. *Educational Psychology Review, 20*(4), 373–389.

Freislich, M.-R., & Bowen-James, A. (2001). Gender, attribution and success in tertiary mathematics. In J. Bobis, B. Perry, & M. Mitchelmore (Eds.), *Numeracy and beyond (Proceedings of the 24th annual conference of the Mathematics Education Research Group of Australasia)* (pp. 231–237). Sydney: MERGA.

Freudenthal, H. (1991). *Revisiting mathematics education: China Lectures.* Dordrecht and Boston: Kluwer Academic Publishers.

Frey, A., Asseburg, R., Carstensen, C. H., Ehmke, T., & Blum, W. (2007). Mathematische Kompetenz. In M. Prenzel et al. (Eds.), *PISA 2006* (pp. 249–276). Münster: Waxmann.

Frey, A., Asseburg, R., Ehmke, T., & Blum, W. (2008). Mathematische Kompetenz im Ländervergleich. In M. Prenzel, C. Artelt, J. Baumert, W. Blum, M. Hammann, & E. Klieme (Eds.), *PISA 2006 in Deutschland. Die Kompetenzen der Jugendlichen im dritten Ländervergleich* (pp. 127–147). Münster: Waxmann.

Glasmachers, E., Griese, B., Kallweit, M., & Roesken, B. (2011). Supporting engineering students in mathematics. In B. Ubuz (Ed.), *Proceedings of the 35th Conference of the International Group for the Psychology of Mathematics Education* (Vol. 1, p. 304). Ankara, Turkey: PME.

Goldin, G. A. (2002). Affect, meta-affect, and mathematical belief structures. In G. Leder, E. Pehkonen, & G. Törner (Eds.), *Beliefs* (Vol. 31, pp. 59–72). Dordrecht and Boston: Kluwer Academic Publishers.

Göller, R., Kortemeyer, J., Liebendörfer, M., Biehler, R., Hochmuth, R., Krämer, J., . . . Schreiber, S. (2013). Instrumentenentwicklung zur Messung von Lernstrategien in mathematikhaltigen Studiengängen. In G. Greefrath, F. Käpnick, & M. Stein (Eds.), *Beiträge zum Mathematikunterricht 2013* (pp. 360–363). Münster and Germany: WTM.

Gómez-Chacón, I. M., García-Madruga, J. A., Vila, J. O., Elosúa, M. R., & Rodríguez, R. (2014). The dual processes hypothesis in mathematics performance: beliefs, cognitive reflection, reasoning and working memory. *Learning and Individual Differences*, *29*, 67–73.

Gravemeijer, K., & Cobb, P. (2006). Design research from a learning design perspective. In J. van den Akker, K. Gravemeijer, S. McKenney, & N. Nieveen (Eds.), *Educational design research* (pp. 45–85). London, New York: Routledge.

Gravemeijer, K., & Cobb, P. (2013). Design research from a learning design perspective. In T. Plomp & N. Nieveen (Eds.), *Educational design research, part A: An introduction* (pp. 72–113). Enschede, Netherlands: Netherlands Institute for Curriculum Development (SLO).

Gray, E. M., & Tall, D. O. (1991). Duality, ambiguity and flexibility in successful mathematical thinking. In F. Furinghetti (Ed.), *Proceedings of the 15th International Conference for the Psychology of Mathematics Education* (Vol. 2, pp. 72–79). Genoa and Italy: University of Genoa.

Gray, E. M., & Tall, D. O. (1994). Duality, ambiguity and flexibility: A proceptual view of simple arithmetic. *The Journal for Research in Mathematics Education, 26*(2), 115–141.

Gray, E. M., & Tall, D. O. (2001). Relationships between embodied objects and symbolic procepts: An explanatory theory of success and failure in mathematics. In M. van den Heuvel-Panhuizen (Ed.), *Proceedings of the 25th Conference of the International Group for the Psychology of Mathematics Education, 3* (Vol. 3, pp. 65–72). Utrecht, Netherlands: PME.

Greefrath, G., & Hoever, G. (2016). Was bewirken Mathematik-Vorkurse? Eine Untersuchung zum Studienerfolg nach Vorkursteilnahme an der FH Aachen. In A. Hoppenbrock, R. Biehler, R. Hochmuth, & H.-G. Rück (Eds.), *Lehren und Lernen von Mathematik in der Studieneingangsphase* (pp. 517–530). Wiesbaden: Springer Spektrum.

Griese, B., Glasmachers, E., Kallweit, M., & Roesken, B. (2011). Engineering students and their learning of mathematics. In B. Roesken & M. Casper (Eds.), *Current State of Research on Mathematical Beliefs XVII* (pp. 85–96). Bochum: Professional School of Education, RUB.

Griese, B., & Kallweit, M. (2014). Lerntagebücher in der Studieneingangsphase - eine Bilanz. In J. Roth & J. Ames (Eds.), *Beiträge zum Mathematikunterricht 2014* (pp. 455–458). Münster: WTM.

Griese, B., & Kallweit, M. (2016). Lernverhalten und Klausurerfolg in der Ingenieurmathematik - Selbsteinschätzung und Dozentensicht. In Institut für Mathematik und Informatik Heidelberg (Ed.), *Beiträge zum Mathematikunterricht 2016.* Münster: WTM.

Griese, B., Lehmann, M., & Roesken-Winter, B. (2015). Refining questionnaire-based assessment of STEM students' learning strategies. *International Journal of STEM Education, 2:12.* Retrieved 16.08.2015, from www .stemeducationjournal.com/content/pdf/s40594-015-0025-9.pdf

Grüßing, M. (2009). Mathematische Kompetenzentwicklung zwischen Elementar- und Primarbereich: Zusammenfassung und Forschungsdesiderata. In A. Heinze & M. Grüßing (Eds.), *Mathematiklernen vom Kindergarten bis zum Studium* (pp. 53–58). Münster: Waxmann.

Gueudet, G. (2008). Investigating the secondary–tertiary transition. *Educational Studies in Mathematics, 67*(3), 237–254.

Halai, A. (2014). Social justice through mathematics education: Skilling youth for societal participation. In P. Liljedahl, C. Nicol, S. Oesterle, & D. Allan (Eds.), *Proceedings of the 38th Conference of the International Group for the Psychology of Mathematics Education and the 36th Conference of the North American Chapter of the Psychology of Mathematics Education* (Vol. Vol. 1, pp. 67–71). PME.

Hannula, M. S., Evans, J., Philippou, G., & Zan, R. (2004). *Affect in mathematics education - exploring theoretical frameworks.* Retrieved 09.02.2016, from http://www.emis.de/proceedings/PME28/RF/RF001.pdf

Hannula, M. S., Maijala, H., & Pehkonen, E. (2004). Development of understanding and self-confidence in mathematics; grades 5-8. In M. J. Hoines & A. B. Fuglestad (Eds.), *Proceedings of the 28th Conference of the International Group for the Psychology of Mathematics Education.* (Vol. 3, pp. 17–24). Bergen, Norway: PME.

Härterich, J., Kiss, C., Rooch, A., Mönnigmann, M., Schulze Darup, M., & Span, R. (2012). MathePraxis – Connecting first-year mathematics with engineering applications. *European Journal of Engineering Education, 37*(3), 255–266. doi: 10.1080/03043797.2012.681295

Härterich, J., & Rooch, A. (Eds.). (2013). *Das Mathe-Praxis-Buch - Wie Ingenieure Mathematik anwenden: Projekte für die Bachelor-Phase.* Wiesbaden: Springer Vieweg.

Hausberger, T. (2013). On the concept of (homo)morphism: A key notion in the learning of linear algebra. In B. Ubuz, C. Haser, & M. A. Mariotti (Eds.), *Proceedings of the Eighth Conference of the European Society for Research in Mathematics Education* (pp. 2346–2355). Ankara: Middle East Technical University.

Heckhausen, H. (1977). Achievement motivation and its constructs: A cognitive model. *Motivation and Emotion, 1*(4), 283–329.

Heckhausen, H. (1989). *Motivation und Handeln* (2nd ed.). Berlin and Heidelberg, New York: Springer.

Heckhausen, H., & Heckhausen, J. (2010). *Motivation und Handeln* (4th ed.). Berlin [u.a.]: Springer.

Heiss, C., & Embacher, F. (2016). Effizienz von Mathematik-Vorkursen an der Fachhochschule Technikum Wien - ein datengestützter Reflexionsprozess. In A. Hoppenbrock, R. Biehler, R. Hochmuth, & H.-G. Rück (Eds.), *Lehren und Lernen von Mathematik in der Studieneingangsphase* (pp. 277–293). Wiesbaden: Springer Spektrum.

Henn, G., & Polaczek, C. (2007). Studienerfolg in den Ingenieurwissenschaften. *Das Hochschulwesen, 55*(5), 144–147.

Henn, H.-W. (2003). Working and learning in the real world: A mathematics education project in Baden-Wuerttemberg. In S. J. Lamon, W. A. Parker, & K. Houston (Eds.), *Mathematical modelling* (pp. 71–79). Chichester: Horwood Publishing Ltd.

Henn, H.-W., & Kaiser, G. (2001). Mathematik – ein polarisierendes Schulfach. *Zeitschrift für Erziehungswissenschaft, 4*(3), 359–380.

Hershkowitz, R., Schwarz, B. B., & Dreyfus, T. (2001). Abstraction in context: Epistemic actions. *Journal of Research in Mathematics Education, 32*, 195–222.

Hetze, P. (2011). *Nachhaltige Hochschulstrategien für mehr MINT-Absolventen.* Essen: Edition Stifterverband.

Heublein, U., & Barthelmes, T. (2010). Woran Studierende scheitern - Die Studienstrukturreform führt zu einer Verschiebung bei den Ursachen für einen Studienabbruch. *HIS Magazin*(2), 5–7.

Heublein, U., Richter, J., Schmelzer, R., & Sommer, D. (2012). *Die Entwicklung der Schwund- und Studienabbruchquoten an den deutschen Hochschulen: Statistische Berechnungen auf der Basis des Absolventenjahrgangs 2010* (Vol. 3). Hannover: HIS.

Heublein, U., Richter, J., Schmelzer, R., & Sommer, D. (2014). *Die Entwicklung der Studienabbruchquoten an den deutschen Hochschulen: Statistische Berechnungen auf der Basis des Absolventenjahrgangs 2012.* Hannover: Deutsches Zentrum für Hochschul- und Wissenschaftsforschung.

Heublein, U., Schmelzer, R., & Sommer, D. (2008). *Die Entwicklung der Studienabbruchquote an den deutschen Hochschulen: Ergebnisse einer Berechnung des Studienabbruchs auf der Basis des Absolventenjahrgangs 2006.* Hannover: HIS.

Hilpert, J., Stempien, J., van der Hoeven Kraft, K. J., & Husman, J. (2013). Evidence for the latent factor structure of the MSLQ: A new conceptualization of an established questionnaire. *SAGE-Open, 3*, 1–10.

Hochmuth, R., Roesken-Winter, B., & Jaworski, B. (2013). Engineering students' learning of mathematics: Addressing mathematical competencies. In A. M. Lindmeier & A. Heinze (Eds.), *Proceedings of the 37th Conference of the International Group for the Psychology of Mathematics Education.* (Vol. 1, p. 217). Kiel: PME.

Hoffkamp, A., Paravicini, W., & Schnieder, J. (2016). Denk- und Arbeitsstrategien für das Lernen von Mathematik am Übergang Schule-Hochschule. In A. Hoppenbrock, R. Biehler, R. Hochmuth, & H.-G. Rück (Eds.), *Lehren und Lernen von Mathematik in der Studieneingangsphase* (pp. 295–309). Wiesbaden: Springer Spektrum.

Hoffkamp, A., Schnieder, J., & Paravicini, W. (2013). Mathematical enculturation, argumentation and proof at the transition from school to university. In B. Ubuz, C. Haser, & M. A. Mariotti (Eds.), *Proceedings of the Eighth Conference of the European Society for Research in Mathematics Education* (pp. 2356–2365). Ankara: Middle East Technical University.

Hoppenbrock, A., Biehler, R., Hochmuth, R., & Rück, H.-G. (Eds.). (2016). *Lehren und Lernen von Mathematik in der Studieneingangsphase: Herausforderungen und Lösungsansätze.* Wiesbaden: Springer Spektrum.

Hoyles, C., Newman, K., & Noss, R. (2001). Changing patterns of transition from school to university mathematics. *International Journal of Mathematics Education in Science and Technology*, *32*(6), 829–845.

Hyde, J. S., Fennema, E., Ryan, M., Frost, L. A., & Hopp, C. (1990). Gender comparisons of mathematics attitudes and affect. *Psychology of Women Quarterly*, *14*(3), 299–324.

Jaworski, B. (2008). Helping engineers learn mathematics: A developmental research approach. *Teaching Mathematics and its Applications*, *27*(3), 160–166.

Jaworski, B., Treffert-Thomas, S., & Bartsch, T. (2009). Characterizing the teaching of university mathematics: A case of linear algebra. In M. Tzekaki, M. Kaldrimidou, & H. Sakonidis (Eds.), *Proceedings of the 33rd Conference of the International Group for the Psychology of Mathematics Education* (Vol. 3, pp. 249–256). Thessaloniki: PME.

Juter, K. (2005). Students' attitudes to mathematics and performance in limits of functions. *Mathematics Education Research Journal*, *17*(2), 91–110.

Kallweit, M., & Griese, B. (2014). Serious Gaming an der Hochschule - Mit Avataren zum Studienerfolg? In J. Roth & J. Ames (Eds.), *Beiträge zum Mathematikunterricht 2014* (pp. 591–594). Münster: WTM.

Kelly, A. E. (2006). Quality criteria for design research: Evidence and commitments. In J. van den Akker, K. Gravemeijer, S. McKenney, & N. Nieveen (Eds.), *Educational design research* (pp. 107–118). London, New York: Routledge.

Khiat, H. (2010). A grounded theory approach: Conceptions of understanding in engineering mathematics learning. *The Qualitative Report*, *15*(6), 1459–1488.

Kiss, T., & Kallweit, M. (2015). Der MathePlus Companion - digitale Unterstützung zur Lernstrukturierung. In F. Caluori, H. Linneweber-Lammerskitten, & C. Streit (Eds.), *Beiträge zum Mathematikunterricht 2015*. Münster: WTM.

Kline, P. (1994). *An easy guide to factor analysis*. London and New York: Routledge.

Kloosterman, P. (1988). Self-confidence and motivation in mathematics. *Journal of Educational Psychology*, *80*(3), 345–351.

KMK. (2012). *Bildungsstandards im Fach Mathematik für die Allgemeine Hochschulreife: (Beschluss der Kultusministerkonferenz vom 18.10.2012)*. Retrieved 12.10.2015, from http://www.kmk.org/fileadmin/veroeffentlichungen_beschluesse/ 2012/2012_10_18-Bildungsstandards-Mathe-Abi.pdf

Knight, D. W., Carlson, L. E., & Sullivan, J. F. (Eds.). (2007). *Improving engineering student retention through hands-on, team based, first-year design projects*.

Knospe, H. (2013). Zehn Jahre Eingangstest Mathematik an Fachhochschulen in Nordrhein-Westfalen. In D. Schott, M. Primbs, & J. Vorloeper (Eds.), *Mathematik in ingenieurwissenschaftlichen Studiengängen, Proceedings zum 10. Workshop* (pp. 19–24). Mülheim: Hochschule Ruhr West.

Koch, K. (2006). Der Übergang von der Grundschule in die weiterführende Schule als biographische und pädagogische Herausforderung. In A. Ittel, L. Stecher, H. Merkens, & J. Zinnecker (Eds.), *Jahrbuch Jugendforschung* (pp. 69–89). Wiesbaden: VS Verlag für Sozialwissenschaften.

Köller, O., Baumert, J., & Bos, W. (2001). TIMSS. Third International Mathematics and Science Study. Dritte Internationale Mathematik- und Naturwissenschaftsstudie. In F. E. Weinert (Ed.), *Leistungsmessung in Schulen* (pp. 269–284). Weinheim: Beltz.

Köller, O., & Schiefele, U. (2003). Selbstreguliertes Lernen im Kontext von Schule und Hochschule: Editorial zum Themenschwerpunkt. *Zeitschrift für Pädagogische Psychologie / German Journal of Educational Psychology, 17*(3/4), 155–157.

Kürten, R., Greefrath, G., Harth, T., & Pott-Langemeyer, M. (2014). Die Rechenbrücke - ein fachbereichübergreifendes Foschungs- und Entwicklungsprojekt. *Zeitschrift für Hochschulentwicklung, 9*(4), 17–38.

Lam, T. C., & Bengo, P. (2003). A comparison of three retrospective self-reporting methods of measuring change in instructional practice. *American Journal of Evaluation, 223*(1), 3–13.

Landmann, M., & Schmitz, B. (2007a). Die Kombination von Trainings mit standardisierten Tagebüchern: Angeleitete Selbstbeobachtung als Möglichkeit der Unterstützung von Trainingsmaßnahmen. In M. Landmann & B. Schmitz (Eds.), *Selbstregulation erfolgreich fördern* (pp. 151–163). Stuttgart: Kohlhammer.

Landmann, M., & Schmitz, B. (Eds.). (2007b). *Selbstregulation erfolgreich fördern: Praxisnahe Trainingsprogramme für effektives Lernen.* Stuttgart: Kohlhammer.

Larcombe, A. (1985). *Mathematical learning difficulties in the secondary school: Pupil needs and teacher roles.* Milton Keynes and Philadelphia: Open University Press.

Lazarus, R. S. (1991). *Emotion and adaptation.* New York: Oxford University Press.

Leder, G. (1995). Equity inside the mathematics classroom: Fact or artifact? In W. G. Secada, E. Fennema, & L. Byrd Adajian (Eds.), *New directions for equity in mathematics education* (pp. 209–224). Cambridge and New York: Cambridge University Press.

Leder, G., & Grootenboer, P. (2005). Affect and mathematics education. *Mathematics Education Research Journal, 17*(2), 1–8.

Leder, G., Pehkonen, E., & Törner, G. (Eds.). (2002). *Beliefs: A hidden variable in mathematics education?* (Vol. 31). Dordrecht and Boston: Kluwer Academic Publishers.

Leongson, J. A., & Limjap, A. A. (2003). *Assessing the mathematics achievement of college freshmen using Piaget's logical operations.* Retrieved 01.09.2013, from www.cimt.plymouth.ac.uk/journal/limjap.pdf

Lester, F. K., Garofalo, J., & Kroll, D. L. (1989). Self-confidence, interest, beliefs, and metacognition: Key influences on problem-solving behavior. In D. B. McLeod & V. M. Adams (Eds.), *Affect and mathematical problem solving: A new perspective* (pp. 75–88). New York: Springer.

Liebendörfer, M., Hochmuth, R., Schreiber, S., Göller, R., Kolter, J. B. R., Kortemeyer, J., & Ostsieker, L. (2014). Vorstellung eines Fragebogens zur Erfassung von Lernstrategien in mathematikhaltigen Studiengängen. In J. Roth & J. Ames (Eds.), *Beiträge zum Mathematikunterricht 2014* (pp. 739–742). Münster: WTM.

Liston, M., & O'Donoghue, J. (2008). *The influence of affective variables on students' transition to university mathematics.* Retrieved from http://tsg.icme11.org/tsg/show/31

Livingston, J. A. (2003). Metacognition: An overview. *Psychology, 13,* 259–266.

Lovelace, M., & Brickman, P. (2013). Best practices for measuring students' attitudes towards learning science. *CBE Life Sciences Education, 12,* 606–617.

Ma, X., & Kishor, N. (1997). Assessing the relationship between attitude toward mathematics and achievement in mathematics: A meta-analysis. *Journal for Research in Mathematics Education, 28*(1), 26–47.

Maaz, K., Baumert, J., Gresch, C., & McElvany, N. (Eds.). (2010). *Der Übergang von der Grundschule in die weiterführende Schule: Leistungsgerechtigkeit und regionale, soziale und ethnisch-kulturelle Disparitäten.* Bonn, Berlin: BMBF.

Maaz, K., Gresch, C., McElvany, N., Jonkmann, K., & Baumert, J. (2010). Theoretische Konzepte für die Analyse von Bildungsübergängen: Adaptation ausgewählter Ansätze für den Übergang von der Grundschule in die weiterführenden Schulen des Sekundarschulsystems. In K. Maaz, J. Baumert, C. Gresch, & N. McElvany (Eds.), *Der Übergang von der Grundschule in die weiterführende Schule: Leistungsgerechtigkeit und regionale, soziale und ethnisch-kulturelle Disparitäten* (pp. 64–85). Bonn, Berlin: BMBF.

Mandler, G. (1984). *Mind and body: Psychology of emotion and stress*. New York: W.W. Norton.

McGee, C., Ward, R., Gibbons, J., & Harlow, A. (2003). *Transition to secondary school: A literature review*. Hamilton, NZ: University of Waikato. Retrieved 27.10.2015, from https://www.educationcounts.govt.nz/publications/schooling/5431

McKenney, S. E., Nieveen, N., & van den Akker, J. (2006). Design research from a curriculum perspective. In J. van den Akker, K. Gravemeijer, S. McKenney, & N. Nieveen (Eds.), *Educational design research* (pp. 67–90). London, New York: Routledge.

McKenney, S. E., & Reeves, T. C. (2012). *Conducting educational design research*. London and New York: Routledge.

McKenney, S. E., & Reeves, T. C. (2013). Systematic review of design-based research progress: Is a little knowledge a dangerous thing? *Educational Researcher*, *42*(2), 97–100.

McLeod, D. B. (1992). Research on affect in mathematics education: A reconceptualization. In D. A. Grouws (Ed.), *Handbook of research on mathematics teaching and learning* (pp. 575–596). New York: Macmillan.

Meyberg, K., & Vachenauer, P. (2003). *Höhere Mathematik 1: Differential- und Integralrechnung, Vektor- und Matrizenrechnung* (6th ed.). Berlin and Heidelberg: Springer.

Moore, R. C. (1994). Making the transition to formal proof. *Educational Studies in Mathematics*, *27*, 249–266.

Morgan, W. G. (2002). Origin and history of the earliest Thematic Apperception Test pictures. *Journal of personality assessment*, *79*(3), 422–445.

Mündemann, F., Fröhlich, S., Ioffe, O. B., & Krebs, F. (2016). Kompetenzbrücken zwischen Schule und Hochschule. In A. Hoppenbrock, R. Biehler, R. Hochmuth, & H.-G. Rück (Eds.), *Lehren und Lernen von Mathematik in der Studieneingangsphase* (pp. 321–338). Wiesbaden: Springer Spektrum.

Mungenast, S. (2015). Zur Bedeutung von Metakognition beim Lehren und Lernen von Mathematik - Entwicklung eines Kategoriensystems. In F. Caluori, H. Linneweber-Lammerskitten, & C. Streit (Eds.), *Beiträge zum Mathematikunterricht 2015* (pp. 648–651). Münster: WTM.

Murray, H. A. (1943). *Thematic Apperception Test manual*. Cambridge, MA: Harvard University Press.

Nardi, E., & Steward, S. (2002). I could be the best mathematician in the world ... if I actually enjoyed it. *Mathematics Teaching*, *179*, 41–44.

Nenniger, P. (1999). On the role of motivation in self-directed learning. The 'two-shells-model of motivated self-directed learning' as a structural explanatory concept. *European Journal of Psychology of Education*, *14*(1), 71–86.

Nerdinger, F. W. (2014). Motivierung. In H. Schuler & U. P. Kanning (Eds.), *Lehrbuch der Personalpsychologie* (pp. 725–764). Göttingen: Hogrefe.

Neumann, I., Rösken-Winter, B., Lehmann, M., Duchhardt, C., Heinze, A., & Nickolaus, R. (2015). Measuring mathematical competences of engineering students at the beginning of their studies. *Peabody Journal of Education*, *90*(4), 465–476.

Nì Fhloinn, E., Fitzmaurice, O., Mac an Bhaird, C., & O'Sullivan, C. (2014). Student perception of the impact of mathematics support in higher education. *International Journal of Mathematics Education in Science and Technology*, *45*(7), 953–967.

Nieveen, N. (1999). Prototyping to reach product quality. In J. van den Akker, R. M. Branch, K. Gustafson, N. Nieveen, & T. Plomp (Eds.), *Design approaches and tools in education and training* (pp. 125–136). Boston: Kluwer Academic Publishers.

Nieveen, N., McKenney, S., & van den Akker, J. (2006). Educational design research: The value of variety. In J. van den Akker, K. Gravemeijer, S. McKenney, & N. Nieveen (Eds.), *Educational design research* (pp. 151–158). London, New York: Routledge.

Nimon, K., Zigarni, D., & Allen, J. (2011). Measures of program effectiveness based on retrospective pretest data: Are all created equal? *American Journal of Evaluation*, *32*(1), 8–28.

Niss, M. (2003). Mathematical competencies and the learning of mathematics: The Danish KOM project. In a. Gagatsis & S. Papastravidis (Eds.), *3rd Mediterranean Conference on Mathematics Education* (pp. 115–124). Athens, Greece: Hellenic Mathematical Society and Cyprus Mathematical Society.

Noss, R., & Kent, P. (2002). *The mathematical components of engineering expertise*. Swindon: Economic and Social Research Council.

Papula, L. (2010). *Mathematik für Ingenieure und Naturwissenschaftler - Klausur- und Übungsaufgaben: 632 Aufgaben mit ausführlichen Lösungen zum Selbststudium und zur Prüfungsvorbereitung* (4th ed.). Wiesbaden: Vieweg + Teubner.

Papula, L. (2011). *Mathematik für Ingenieure und Naturwissenschaftler: Band 1* (13th ed.). Wiesbaden: Vieweg+Teubner.

Papula, L. (2012). *Mathematik für Ingenieure und Naturwissenschaftler: Band 2* (13th ed.). Wiesbaden: Vieweg+Teubner.

Pehkonen, E. (1997). Learning results from the viewpoint of equity: Boys, girls and mathematics. *Teaching Mathematics and its Applications, 16*(2), 58–63.

Phillips, D. C. (2006). Assessing the quality of design research proposals: Some philosophical perspectives. In J. van den Akker, K. Gravemeijer, S. McKenney, & N. Nieveen (Eds.), *Educational design research* (pp. 93–99). London, New York: Routledge.

Piaget, J. (1973). *Die Entwicklung der elementaren logischen Strukturen* (1st ed.). Düsseldorf, Germany: Schwann.

Pintrich, P., & de Groot, E. (1990). Motivational and self-regulated learning components of classroom academic performance. *Journal of Educational Psychology, 81*(1), 33–40.

Pintrich, P., Smith, D., Garcia, T., & McKeachie, W. (1993). Reliability and predictive validity of the Motivated Strategies for Learning Questionnaire (MSLQ). *Educational and Psychological Measurement, 53*(3), 801–813.

Plomp, T. (2013). Educational design research: An introduction. In T. Plomp & N. Nieveen (Eds.), *Educational design research, part A: An introduction* (pp. 11–50). Enschede, Netherlands: Netherlands Institute for Curriculum Development (SLO).

Plomp, T., & Nieveen, N. (Eds.). (2013a). *Educational design research, part A: An introduction.* Enschede, Netherlands: Netherlands Institute for Curriculum Development (SLO).

Plomp, T., & Nieveen, N. (Eds.). (2013b). *Educational design research, part B: Illustrative cases.* Enschede, Netherlands: Netherlands Institute for Curriculum Development (SLO).

Prediger, S., Gravemeijer, K., & Confrey, J. (2015). Design research with a focus on learning processes: An overview on achievements and challenges. *ZDM Mathematics Education*(6), 877–891.

Rach, S. (2014). *Individuelle Lernprozesse im Mathematikstudium: Charakteristika mathematischer Lehr-Lern-Prozesse in der Studieneingangsphase und individuelle Bedingungsfaktoren für erfolgreiche Lernprozesse im ersten Semester* (Dissertation). Christian-Albrechts-Universität, Kiel.

Rach, S., & Heinze, A. (2011). Studying mathematics at the university: The influence of learning strategies. In B. Ubuz (Ed.), *Proceedings of the 35th Conference of the International Group for the Psychology of Mathematics Education* (Vol. 4, pp. 9–16). Ankara, Turkey: PME.

Rach, S., & Heinze, A. (2013). Welche Studierenden sind im ersten Semester erfolgreich? Zur Rolle von Selbsterklärungen beim Mathematiklernen in der Studieneingangsphase. *Journal für Mathematik-Didaktik(34)*, 121–147.

Rach, S., Siebert, U., & Heinze, A. (2016). Operationalisierung und empirische Erprobung von Qualitätskriterien für mathematische Lehrveranstaltungen in der Studieneingangsphase. In A. Hoppenbrock, R. Biehler, R. Hochmuth, & H.-G. Rück (Eds.), *Lehren und Lernen von Mathematik in der Studieneingangsphase* (pp. 601–618). Wiesbaden: Springer Spektrum.

Reeves, T. C. (2006). Design research from a technology perspective. In J. van den Akker, K. Gravemeijer, S. McKenney, & N. Nieveen (Eds.), *Educational design research* (pp. 52–66). London, New York: Routledge.

Reiss, K. (2009a). Erwerb mathematischer Kompetenzen in der Sekundarstufe: Zusammenfassung und Forschungsdesiderata. In A. Heinze & M. Grüßing (Eds.), *Mathematiklernen vom Kindergarten bis zum Studium* (pp. 191–198). Münster: Waxmann.

Reiss, K. (2009b). Mathematische Kompetenz zwischen Grundschule und Sekundarstufe: Zusammenfassung und Forschungsdesiderata. In A. Heinze & M. Grüßing (Eds.), *Mathematiklernen vom Kindergarten bis zum Studium* (pp. 117–122). Münster: Waxmann.

Rheinberg, F., Vollmeyer, R., & Burns, B. D. (2001). FAM: Ein Fragebogen zur Erfassung aktueller Motivation in Lern- und Leistungssituationen. *Diagnostica, 47*(2), 57–66.

Roesken, B., & Casper, M. (Eds.). (2011). *Current State of Research on Mathematical Beliefs XVII: Proceedings of the MAVI-17 Conference.* Bochum: Professional School of Education, RUB.

Roesken, B., Hannula, M. S., & Pehkonen, E. (2011). Dimensions of students' views of themselves as learners of mathematics. *ZDM Mathematics Education, 43*(4), 497–506.

Rost, J. (1996). *Lehrbuch Testtheorie, Testkonstruktion.* Bern, Göttingen, Toronto, Seattle: Huber.

Rudolf, M., & Müller, J. (2012). *Multivariate Verfahren: Eine praxisorientierte Einführung mit Anwendungsbeispielen in SPSS ; [Lehrbuch]* (2., überarb. und erw. Aufl. ed.). Göttingen [u.a.]: Hogrefe.

Ruhr-Universität Bochum. (2009a). *Amtliche Bekanntmachungen der Ruhr-Universität Bochum Nr. 808: Satzung zur Änderung der Prüfungsordnung für den Bachelor-Studiengang Maschinenbau und den Master-Studiengang Maschinenbau an der Ruhr-Universität Bochum.* Retrieved 13.08.2013, from http://www.uv.ruhr-uni-bochum.de/dezernat1/amtliche/ ab808.pdf

Ruhr-Universität Bochum. (2009b). *Amtliche Bekanntmachungen der Ruhr-Universität Bochum Nr. 811: Bachelor-/Master- Prüfungsordnung für den Studiengang Bauingenieurwesen an der Ruhr-Universität Bochum.* Retrieved 13.08.2013, from http://www.uv.ruhr-uni-bochum.de/dezernat1/ amtliche/ab811.pdf

Ruhr-Universität Bochum. (2009c). *Amtliche Bekanntmachungen der Ruhr-Universität Bochum Nr. 812: Prüfungsordnung für die Bachelor-/Master-Studiengänge Umwelttechnik und Ressourcenmanagement an der Ruhr-Universität Bochum.* Retrieved 13.08.2013, from www.uv.ruhr-uni -bochum.de/dezernat1/amtliche/ab812.pdf

Ryan, R. M., & Deci, E. L. (2000). Intrinsic and extrinsic motivations: Classic definitions and new directions. *Contemporary Educational Psychology, 25*, 54–67.

Schaub, M., & Bruder, R. (2015). Qualitätskriterien für diagnostische Tests im Übergang Schule - Hochschule. In F. Caluori, H. Linneweber-Lammerskitten, & C. Streit (Eds.), *Beiträge zum Mathematikunterricht 2015* (pp. 1105–1108). Münster: WTM.

Schellings, C. (2011). Applying learning strategy questionnaires: Problems and possibilities. *Metacognition and Learning, 6*(2), 91–109.

Schiefele, U., Streblow, L., Ermgassen, U., & Moschner, B. (2003). Lernmotiva-
tion und Lernstrategien als Bedingungen der Studienleistung: Ergebnisse
einer Längsschnittstudie / The influence of leanring motivation and learning strategies on college achievement: Results of a longitudinal analysis.
*Zeitschrift für Pädagogische Psychologie / German Journal of Educa-
tional Psychology, 17*(3/4), 185–198.

Schindler, S. (2012). *Aufstiegsangst? Eine Studie zur sozialen Ungleichheit im
historischen Zeitverlauf.* Düsseldorf: Vodafone Stiftung Deutschland.

Schmidt, K., Allgaier, A., Lachner, A., Stucke, B., Rey, S., Frömmel, C., ...
Nückles, M. (2011). Diagnostik und Förderung selbstregulierten Lernens
durch Self-Monitoring-Tagebücher. *Zeitschrift für Hochschulentwicklung,
6*(3), 246–269.

Schmitz, B. (2003). Selbstregulation - Sackgasse oder Weg mit Forschungsper-
spektive? *Zeitschrift für Pädagogische Psychologie, 17*(3/4), 221–232.

Schmitz, B., & Wiese, B. S. (2006). New perspectives for the evaluation of
training sessions in self-regulated learning: Time-series analysis of diary
data. *Contemporary Educational Psychology, 31*, 64–69.

Schmitz, M., & Grünberg, K. (2016). Erfahrungen aus der Mathe-Klinik. In
A. Hoppenbrock, R. Biehler, R. Hochmuth, & H.-G. Rück (Eds.), *Lehren
und Lernen von Mathematik in der Studieneingangsphase* (pp. 451–463).
Wiesbaden: Springer Spektrum.

Schoenfeld, A. (1992). Learning to think mathematically: Problem solving,
metacognition and sense-making in mathematics. In D. A. Grouws (Ed.),
Handbook of research on mathematics teaching and learning (pp. 334–
370). New York: Macmillan.

Schüler, J., & Engeser, S. (2009). Incentives and flow experiences in learning
settings and the moderating role of individual differences. In M. Wos-
nitza, S. Karabenick, A. Efklides, & P. Nenniger (Eds.), *Contemporary
motivation research* (pp. 339–358). Göttingen, New York: Hogrefe &
Huber.

Schwarz, B. B., Dreyfus, T., & Hershkowitz, R. (2009). The nested epistemic
actions model for abstraction in context. In B. B. Schwarz, T. Dreyfus, &
R. Hershkowitz (Eds.), *Transformation of knowledge through classroom
interaction* (pp. 11–41). London and New York: Routledge.

Selden, A., & Selden, J. (2005). Perspectives on advanced mathematical
thinking. *Mathematical Thinking and Learning, 7*(1), 1–13.

Shavelson, R. J., & Towne, L. (2002). *Scientific research in education*. Washington, DC: National Academy Press.

Skinner, B. F. (2002). *Beyond freedom and dignity*. Indianapolis, Ind.: Hackett Pub.

Skovsmose, O., Valero, P., & Christensen, O. R. (Eds.). (2009). *University science and mathematics education in transition*. Boston, MA: Springer US.

Snow, R. E., & Farr, M. J. (Eds.). (1987). *Aptitude, learning and instruction, Volume 3: Conative and affective process analyses*. Hillsdale, NJ and London: Lawrence Erlbaum Associates.

Stavy, R., & Tirosh, D. (2000). *How students (mis-)understand science and mathematics: Intuitive rules*. New York: Teachers College Press.

Stern, E. (1997). Early training: Who, what, when, why, and how? In M. Beishuizen, K. Gravemeijer, & E. van Lieshout (Eds.), *The role of contexts and models of mathematical strategies and procedures* (pp. 239–253). Culembourg, NL: Technipress.

Szczyrba, B., & Wiemer, M. (2011). Forschungsfeld Tutorien: vom Nachhilfebetrieb zum Motor guter Lehre an Hochschulen. *Zeitschrift für Hochschulentwicklung, 6*(3), 165–170.

Tait, H., Entwistle, N., & McCune, V. (1998). ASSIST: A reconceptualisation of the approaches to studying inventory. In C. Rust (Ed.), *Improving student learning: Improving students as learners* (pp. 262–271). Oxford: Oxford Centre for Staff and Learning Development.

Tall, D. O. (Ed.). (1991a). *Advanced mathematical thinking* (Vol. 11). Dordrecht and Boston: Kluwer Academic Publishers.

Tall, D. O. (1991b). Reflections. In D. O. Tall (Ed.), *Advanced mathematical thinking* (Vol. 11, pp. 251–259). Dordrecht and Boston: Kluwer Academic Publishers.

Tall, D. O. (1999). Reflections on APOS theory in elementary and advanced mathematical thinking. In O. Zaslavsky (Ed.), *Proceedings of the 23rd Conference of the International Group of Psychology in Mathematics Education, Vol. 1* (Vol. 1, pp. 111–118). Haifa, Israel.

Tall, D. O. (2004). Building theories: The three worlds of mathematics. *For the Learning of Mathematics, 24*(1), 29–33. Retrieved from http://flm-journal.org/index.php?do=show&lang=en&vol=24&num=1

Tall, D. O., & Vinner, S. (1981). Concept image and concept definition in mathematics with particular reference to limits and continuity. *Educational Studies in Mathematics*, *12*, 151–169.

Trapmann, S., Hell, B., Weigand, S., & Schuler, H. (2007). Die Validität von Schulnoten zur Vorhersage des Studienerfolgs - eine Metaanalyse. *Zeitschrift für Pädagogische Psychologie*, *21*(1), 11–27.

van den Akker, J. (1999). Principles and methods of development research. In J. van den Akker, R. M. Branch, K. Gustafson, N. Nieveen, & T. Plomp (Eds.), *Design approaches and tools in education and training* (pp. 1–14). Boston: Kluwer Academic Publishers.

van den Akker, J. (2013). Curricular development research as a specimen of educational design research. In T. Plomp & N. Nieveen (Eds.), *Educational design research, part A: An introduction* (pp. 52–71). Enschede, Netherlands: Netherlands Institute for Curriculum Development (SLO).

van den Akker, J., Branch, R. M., Gustafson, K., Nieveen, N., & Plomp, T. (Eds.). (1999). *Design approaches and tools in education and training*. Boston: Kluwer Academic Publishers.

van den Akker, J., Gravemeijer, K., McKenney, S., & Nieveen, N. (2006). Introducing educational design research. In J. van den Akker, K. Gravemeijer, S. McKenney, & N. Nieveen (Eds.), *Educational design research* (pp. 3–7). London, New York: Routledge.

Vester, F. (2004). *Denken, Lernen, Vergessen: Was geht in unserem Kopf vor, wie lernt das Gehirn und wann läßt es uns im Stich?* (30th ed.). München: DTV.

Viirmann, O. (2013). What we talk about when we talk about functions - characteristics of the function concept in the discursive practice of three university teachers. In B. Ubuz, C. Haser, & M. A. Mariotti (Eds.), *Proceedings of the Eighth Conference of the European Society for Research in Mathematics Education* (pp. 2466–2475). Ankara: Middle East Technical University.

Vinner, S. (1991). The role of definitions in the teaching and learning of mathematics. In D. O. Tall (Ed.), *Advanced mathematical thinking* (Vol. 11, pp. 65–81). Dordrecht and Boston: Kluwer Academic Publishers.

Vohs, K. D., & Baumeister, R. F. (Eds.). (2011). *Handbook of self-regulation: Research, Theory, and Applications* (2nd ed.). New York: Guilford Press.

von Glasersfeld, E. (1991). Questions and answers about radical constructivism. In M. K. Pearsall (Ed.), *Scope, sequence, and coordination of secondary school science, Vol. II* (pp. 169–182). Washington, D.C.: The National Science Teachers Association.

Vygotsky, L. S. (1978). *Mind in society: The Development of Higher Psychological Processes*. Cambridge: Harvard University Press.

Walker, D. (2006). Toward productive design studies. In J. van den Akker, K. Gravemeijer, S. McKenney, & N. Nieveen (Eds.), *Educational design research* (pp. 8–13). London, New York: Routledge.

Watson, J. B. (1913). Psychology as the behaviorist views it. *Psychological Review(20)*, 158–177.

Weiner, B. (1994). *Motivationspsychologie* (3rd ed.). Weinheim: Beltz, Psychologie-Verlag-Union.

Weinhold, C. (2014). Wiederholungs- und Unterstützungskurse in Mathematik für Ingenieurwissenschaften an der TU Braunschweig. In I. Bausch et al. (Eds.), *Mathematische Vor- und Brückenkurse* (pp. 241–255). Wiesbaden: Springer Spektrum.

Weinstein, C. E., & Mayer, R. E. (1986). The teaching of learning strategies. In M. C. Wittrock (Ed.), *Handbook of research on teaching* (pp. 315–327). New York: Macmillan.

Weinstein, C. E., & Palmer, D. R. (2002). *Learning and Study Strategies Inventory (LASSI): User's manual* (2nd ed.). Clearwater, FL: H&H Publishing.

Weinstein, C. E., Schulte, A. C., & Palmer, D. R. (1987). *Learning and Study Strategies Inventory (LASSI)*. Clearwater, FL: H & H Publishing.

Wild, K.-P. (1994). *Lernstrategien im Studium: Ergebnisse zur Faktorenstruktur und Reliabilität eines neuen Fragebogens*. Universität Potsdam and Humanwissenschaftliche Fakultät, Institut für Psychologie.

Wild, K.-P. (2000). *Lernstrategien im Studium: Strukturen und Bedingungen*. Münster: Waxmann.

Wild, K.-P. (2005). Individuelle Lernstrategien von Studierenden. Konsequenzen für die Hochschuldidaktik und die Hochschullehre. *Beiträge zur Lehrerbildung, 23*(2), 191–206.

Wild, K.-P., & Schiefele, U. (1994). Lernstrategien im Studium. Ergebnisse zur Faktorenstruktur und Reliabilität eines neuen Fragebogens. *Zeitschrift für Differentielle und Diagnostische Psychologie, 15*, 185–200.

Winne, P. H. (1996). A metacognitive view of individual differences in self-regulated learning. *Learning and Individual Differences, 8*, 327–353.

Winsløw, C. (2013). The transition from university to high school and the case of exponential functions. In B. Ubuz, C. Haser, & M. A. Mariotti (Eds.), *Proceedings of the Eighth Conference of the European Society for Research in Mathematics Education* (pp. 2476–2485). Ankara: Middle East Technical University.

Winter, H. (1996). Mathematikunterricht und Allgemeinbildung. *Mitteilungen der Gesellschaft für Didaktik der Mathematik, 61*, 37–46.

Wood, T., Williams, G., & McNeal, B. (2006). Children's mathematical thinking in different classroom cultures. *Journal for Research in Mathematics Education, 37*, 222–255.

Yackel, E., & Cobb, P. (1996). Sociomathematical norms, argumentation, and autonomy in mathematics. *Journal for Research in Mathematics Education, 27*(4), 458–477.

Zan, R., Brown, L., Evans, J., & Hannula, M. S. (2006). Affect in mathematics education: An introduction. *Educational Studies in Mathematics, 63*, 113–121.

Zimmerman, B. J. (1989). A social cognitive view of self-regulated academic learning. *Journal of Educational Psychology, 81*, 329–339.

Zimmerman, B. J. (1990). Self-regulated learning and academic achievement: An overview. *Educational Psychologist, 25*(1), 3–17.

Zimmerman, B. J. (2000). Attaining self-regulation: A social cognitive perspective. In M. Boekaerts, P. R. Pintrich, & M. Zeidner (Eds.), *Handbook of self-regulation* (pp. 13–39). San Diego, London: Academic Press.

Zucker, S. (1996). Teaching at the university level. *Notices of the American Mathematical Society, 43*(8), 863–865.

List of Figures

List of Tables

Appendices

A Questionnaires Containing LIST

The attached original questionnaires were used for the pre surveys at the beginning of the semester respectively for the post surveys at the end of the semester in 2012/2013. They contain LIST items in their respective sections titled *Lernverhalten*.

EvaSys	Fragebogen zum Vorlesungsbeginn WS 2012/13	⬛ Electric Paper

Ruhr-Universität Bochum

MP² - Mathe/Plus/Praxis   Mathematik I für MB, BI, UTRM   **MP**²

Markieren Sie so: ☐ ☒ ☐ ☐ ☐ Bitte verwenden Sie einen Kugelschreiber oder nicht zu starken Filzstift. Dieser Fragebogen wird maschinell erfasst.
Korrektur: ☐ ■ ☐ ☒ ☐ Bitte beachten Sie im Interesse einer optimalen Datenerfassung die links gegebenen Hinweise beim Ausfüllen.

1. Informationen

In diesem Jahr findet zum dritten Mal eine Befragung aller Studierenden der Vorlesung Mathematik I bezüglich ihrer Wahl des Studiengangs, ihrer Vorkenntnisse und ihres Lernverhaltens nach dem Übergang an die Universität statt. Im Rahmen des Projektes MP² untersuchen wir, welche Maßnahmen und Aspekte zu einem erfolgreichen Verlauf des ersten Studienjahrs beitragen. Die Erkenntnisse sollen uns helfen, Sie und zukünftige Studierende effektiver zu unterstützen.

Der Zeitaufwand für den gesamten Fragebogen beträgt etwa 15 Minuten.

Wir danken für Ihre Kooperation!

2. Persönliche Angaben

Bitte geben Sie vor Beantwortung der Fragen den unten angegebenen Code ein. Durch diesen wird gewährleistet, dass wir Ihre Angaben **anonym** mit Ihren Antworten anderer Befragungen verknüpfen können. Alle Angaben werden nur innerhalb von MP² verwendet. Sie werden nicht an Dritte außerhalb der Projektgruppe weitergegeben, also insbesondere nicht an Personen, die Ihre Studienleistungen bewerten.

2.1 Erster Buchstabe des eigenen Vornamens (z.B. **Julia** = **J**)

☐ A   ☐ B   ☐ C
☐ D   ☐ E   ☐ F
☐ G   ☐ H   ☐ I
☐ J   ☐ K   ☐ L
☐ M   ☐ N   ☐ O
☐ P   ☐ Q   ☐ R
☐ S   ☐ T   ☐ U
☐ V   ☐ W   ☐ X
☐ Y   ☐ Z

2.2 Erster Buchstabe des eigenen Nachnamens (z.B. **Mustermann** = **M**)

☐ A   ☐ B   ☐ C
☐ D   ☐ E   ☐ F
☐ G   ☐ H   ☐ I
☐ J   ☐ K   ☐ L
☐ M   ☐ N   ☐ O
☐ P   ☐ Q   ☐ R
☐ S   ☐ T   ☐ U
☐ V   ☐ W   ☐ X
☐ Y   ☐ Z

2.3 Dritte Ziffer von rechts der Matrikelnummer   ☐ 0   ☐ 1   ☐ 2
(z.B. 108123456789 = **7**)   ☐ 3   ☐ 4   ☐ 5
                                                     ☐ 6   ☐ 7   ☐ 8
                                                     ☐ 9

2.4 Zweite Ziffer von rechts der   ☐ 0   ☐ 1   ☐ 2
Matrikelnummer (z.B. 108123456789 = **8**)   ☐ 3   ☐ 4   ☐ 5
                                                     ☐ 6   ☐ 7   ☐ 8
                                                     ☐ 9

Figure A.1: Questionnaire Used in the Pre Survey 2012/2013, page 1/6

2. Persönliche Angaben [Fortsetzung]

2.5 Erste Ziffer von rechts der Matrikelnummer (z.B. 10812345678**9** = **9**)

☐ 0 ☐ 1 ☐ 2
☐ 3 ☐ 4 ☐ 5
☐ 6 ☐ 7 ☐ 8
☐ 9

3. Angaben zur Person

3.1 Geschlecht ☐ männlich ☐ weiblich

3.2 Geburtsjahr

3.3 Muttersprache ☐ deutsch ☐ andere

3.4 Lebenssituation ☐ ledig ☐ verheiratet / feste Beziehung ohne Kinder ☐ verheiratet / feste Beziehung mit Kind(ern)

☐ alleinerziehend

3.5 Studiengang ☐ MB ☐ BI ☐ UTRM
☐ Sonstiger

3.6 Fachsemester ☐ 1 ☐ 2 ☐ 3
☐ 4 ☐ 5 oder mehr

3.7 Auf welcher Schulform haben Sie Ihr Hochschulreife erworben? ☐ Gymnasium ☐ Gesamtschule ☐ Berufskolleg
☐ Sonstige

3.8 Abiturjahrgang

3.9 Schul-Mathematik ☐ Leistungskurs ☐ Grundkurs

3.10 Durchschnittliche Note in Mathematik ☐ 1-2 ☐ 3 ☐ 4
☐ nicht ausreichend ☐ keine Angabe

3.11 Schul-Physik / -Technik ☐ Leistungskurs ☐ Grundkurs ☐ weder noch

4. Angaben zum Studium

4.1 Haben Sie den Mathematik-Vorkurs besucht? ☐ ja ☐ nein

4.2 Mit welchen Verkehrsmitteln gelangen Sie zur Ruhr-Universität? (Mehrfachnennungen möglich)
☐ Auto / Motrorrad ☐ Öffentliche Verkehrsmittel ☐ Fahrrad
☐ zu Fuß

4.3 Wie lange brauchen Sie für den Weg zur Ruhr-Universität (in Minuten)? ☐ bis zu 10 ☐ 11 bis 20 ☐ 21 bis 30
☐ 31 bis 45 ☐ 46 bis 60 ☐ 61 bis 120
☐ mehr als 120 Minuten

4.4 Wie finanzieren Sie Ihr Studium? (Mehrfachnennungen möglich)
☐ Eltern ☐ BAFöG ☐ Nebenjob(s)
☐ Stipendium ☐ Sonstiges

4.5 Wenn Sie einem Nebenjob nachgehen, wie viel Zeit wöchentlich (in Stunden) wenden Sie insgesamt dafür auf? ☐ bis zu 5 ☐ mehr als 5, bis zu 10 ☐ mehr als 10, bis zu 15
☐ mehr als 15, bis zu 20 ☐ mehr als 20, bis zu 30 ☐ mehr als 30 Stunden

Questionnaire Used in the Pre Survey 2012/2013, page 2/6

4. Angaben zum Studium   [Fortsetzung]

4.6 Wie viel Zeit wöchentlich (in Stunden) widmen Sie anderen außeruniversitären Aktivitäten, außer Nebenjobs, z.B. für Sport oder andere Hobbys?

☐ bis zu 5   ☐ mehr als 5, bis zu 10   ☐ mehr als 10, bis zu 15

☐ mehr als 15, bis zu 20   ☐ mehr als 20, bis zu 30   ☐ mehr als 30 Stunden

4.7 Wie viel Zeit wöchentlich (in Stunden) bleibt Ihnen fürs Studium, ohne Lehrveranstaltungen und Übungen?

☐ bis zu 5   ☐ mehr als 5, bis zu 10   ☐ mehr als 10, bis zu 15

☐ mehr als 15, bis zu 20   ☐ mehr als 20, bis zu 30   ☐ mehr als 30 Stunden

Wie wichtig waren für Sie die folgenden Gründe bei der Wahl Ihres Studienfaches?

4.8 Persönliche Neigungen und Begabungen	sehr wichtig	☐	☐	☐	☐	sehr unwichtig
4.9 Gute Verdienstmöglichkeiten	sehr wichtig	☐	☐	☐	☐	sehr unwichtig
4.10 Ratschläge von Eltern oder Freunden	sehr wichtig	☐	☐	☐	☐	sehr unwichtig
4.11 Ratschläge von Studien- und Berufsberatern	sehr wichtig	☐	☐	☐	☐	sehr unwichtig
4.12 Generelles Interesse am Fach seit der Schulzeit	sehr wichtig	☐	☐	☐	☐	sehr unwichtig

4.13 Sonstige Gründe

4.14 Welcher dieser Aspekte war der wichtigste?

☐ Persönliche Neigungen und Begabungen   ☐ Gute Verdienstmöglichkeiten   ☐ Ratschläge von Eltern oder Freunden

☐ Ratschläge von Studien- oder Berufsberatern   ☐ Generelles Interesse am Fach seit der Schulzeit   ☐ Sonstige Gründe

4.15 Wann stand Ihre Studienwahl fest?   ☐ zur Schulzeit   ☐ nach dem Abitur

4.16 Trifft folgende Aussage auf Sie zu: Ich wollte ein anderes Fach studieren, habe aber dafür keine Zulassung erhalten?   ☐ Trifft zu   ☐ Trifft nicht zu

4.17 Welche Rolle haben Arbeitsmarktüberlegungen bei Ihrer Studienwahl gespielt?   sehr große Rolle ☐ ☐ ☐ ☐ überhaupt keine Rolle

4.18 Wie gut fühlen Sie sich durch die Schule auf das Studium vorbereitet?   sehr gut ☐ ☐ ☐ ☐ sehr schlecht

5. Lernverhalten

Geben Sie bitte für jede im Folgenden beschriebene Tätigkeit an, wie häufig diese bei Ihnen vorkommt. Sie können Ihre Antworten von **sehr selten** bis **sehr oft** abstufen.

5.1 Ich lerne für mein Studium.   sehr selten ☐ ☐ ☐ ☐ sehr oft

5.2 Ich fertige Tabellen, Diagramme oder Schaubilder an, um den Stoff der Veranstaltung besser strukturiert vorliegen zu haben.   sehr selten ☐ ☐ ☐ ☐ sehr oft

Questionnaire Used in the Pre Survey 2012/2013, page 3/6

5. Lernverhalten [Fortsetzung]

5.3 Ich versuche, Beziehungen zu den Inhalten verwandter Fächer bzw. Lehrveranstaltungen herzustellen. — sehr selten ☐ ☐ ☐ ☐ sehr oft

5.4 Ich präge mir den Lernstoff der Vorlesung durch Wiederholen ein. — sehr selten ☐ ☐ ☐ ☐ sehr oft

5.5 Ich versuche, mir vorher genau zu überlegen, welche Teile eines bestimmten Themengebietes ich lernen muss und welche nicht. — sehr selten ☐ ☐ ☐ ☐ sehr oft

5.6 Wenn ich schwierigen Stoff vorliegen habe, passe ich meine Lerntechnik den höheren Anforderungen an (z. B. durch langsameres Lesen). — sehr selten ☐ ☐ ☐ ☐ sehr oft

5.7 Ich bearbeite Aufgaben zusammen mit meinen Kommilitonen. — sehr selten ☐ ☐ ☐ ☐ sehr oft

5.8 Ich suche nach weiterführendem Material, wenn mir bestimmte Inhalte noch nicht ganz klar sind. — sehr selten ☐ ☐ ☐ ☐ sehr oft

5.9 Wenn ich mir ein bestimmtes Pensum zum Lernen vorgenommen habe, bemühe ich mich, es auch zu schaffen. — sehr selten ☐ ☐ ☐ ☐ sehr oft

5.10 Beim Lernen merke ich, dass meine Gedanken abschweifen. — sehr selten ☐ ☐ ☐ ☐ sehr oft

5.11 Beim Lernen halte ich mich an einen bestimmten Zeitplan. — sehr selten ☐ ☐ ☐ ☐ sehr oft

5.12 Ich lerne an einem Platz, wo ich mich gut auf den Stoff konzentrieren kann. — sehr selten ☐ ☐ ☐ ☐ sehr oft

5.13 Wenn ich während des Lesens nicht alles verstehe, versuche ich, die Lücken festzuhalten und das Material daraufhin noch einmal durchzugehen. — sehr selten ☐ ☐ ☐ ☐ sehr oft

5.14 Ich mache mir kurze Zusammenfassungen der wichtigsten Inhalte als Gedankenstütze. — sehr selten ☐ ☐ ☐ ☐ sehr oft

5.15 Ich nehme mir Zeit, um mit Kommilitonen über den Stoff zu diskutieren. — sehr selten ☐ ☐ ☐ ☐ sehr oft

5.16 Wenn ich einen Fachbegriff nicht verstehe, so schlage ich ihn nach, z.B. in einem Fachbuch oder im Internet. — sehr selten ☐ ☐ ☐ ☐ sehr oft

5.17 Zu neuen Konzepten stelle ich mir praktische Anwendungen vor. — sehr selten ☐ ☐ ☐ ☐ sehr oft

5.18 Ich lese meine Aufzeichnungen / das Skript mehrmals hintereinander durch. — sehr selten ☐ ☐ ☐ ☐ sehr oft

5.19 Ich lege im Vorhinein fest, wie weit ich mit der Durcharbeitung des Stoffes kommen möchte. — sehr selten ☐ ☐ ☐ ☐ sehr oft

5.20 Ich strenge mich auch dann an, wenn mir der Stoff überhaupt nicht liegt. — sehr selten ☐ ☐ ☐ ☐ sehr oft

5.21 Es fällt mir schwer, bei der Sache zu bleiben. — sehr selten ☐ ☐ ☐ ☐ sehr oft

5.22 Ich lege bestimmte Zeiten fest, zu denen ich dann lerne. — sehr selten ☐ ☐ ☐ ☐ sehr oft

5.23 Ich gestalte meine Umgebung so, dass ich möglichst wenig vom Lernen abgelenkt werden. — sehr selten ☐ ☐ ☐ ☐ sehr oft

5.24 Ich gehe meine Aufzeichnungen durch und mache mir dazu eine Gliederung mit den wichtigsten Punkten. — sehr selten ☐ ☐ ☐ ☐ sehr oft

5.25 Ich vergleiche meine Vorlesungsmitschriften mit denen meiner Kommilitonen. — sehr selten ☐ ☐ ☐ ☐ sehr oft

Questionnaire Used in the Pre Survey 2012/2013, page 4/6

5. Lernverhalten [Fortsetzung]

5.26 Fehlende Informationen suche ich mir aus verschiedenen Quellen zusammen (z.B. Internet, Bücher, Fachzeitschriften). sehr selten ☐ ☐ ☐ ☐ sehr oft

5.27 Ich versuche, neue Begriffe oder Theorien auf mir bereits bekannte Begriffe und Theorien zu beziehen. sehr selten ☐ ☐ ☐ ☐ sehr oft

5.28 Ich lerne Schlüsselbegriffe auswendig, um mich in der Prüfung besser an wichtige Inhaltsbereiche erinnern zu können. sehr selten ☐ ☐ ☐ ☐ sehr oft

5.29 Vor dem Lernen eines Stoffgebietes überlege ich mir, wie ich am effektivsten vorgehen kann. sehr selten ☐ ☐ ☐ ☐ sehr oft

5.30 Ich gebe nicht auf, auch wenn der Stoff sehr schwierig oder komplex ist. sehr selten ☐ ☐ ☐ ☐ sehr oft

5.31 Ich ertappe mich dabei, dass ich mit meinen Gedanken ganz woanders bin. sehr selten ☐ ☐ ☐ ☐ sehr oft

5.32 Ich lege die Stunden, die ich täglich mit Lernen verbringe, durch einen Zeitplan fest. sehr selten ☐ ☐ ☐ ☐ sehr oft

5.33 Zum Lernen sitze ich immer am selben Platz. sehr selten ☐ ☐ ☐ ☐ sehr oft

5.34 Ich versuche, den Stoff so zu ordnen, dass ich ihn mir gut einprägen kann. sehr selten ☐ ☐ ☐ ☐ sehr oft

5.35 Ich lasse mich von einem Kommilitonen abfragen und stelle auch ihm Fragen zum Stoff. sehr selten ☐ ☐ ☐ ☐ sehr oft

5.36 Ich ziehe zusätzliche Quellen heran, wenn meine Aufzeichnungen unvollständig sind. sehr selten ☐ ☐ ☐ ☐ sehr oft

5.37 Ich stelle mir manche Sachverhalte bildlich vor. sehr selten ☐ ☐ ☐ ☐ sehr oft

5.38 Ich lerne eine selbst erstellte Übersicht mit den wichtigsten Fachtermini auswendig. sehr selten ☐ ☐ ☐ ☐ sehr oft

5.39 Ich überlege mir vorher, in welcher Reihenfolge ich den Stoff durcharbeite. sehr selten ☐ ☐ ☐ ☐ sehr oft

5.40 Ich lerne auch spätabends und am Wochenende, wenn es sein muss. sehr selten ☐ ☐ ☐ ☐ sehr oft

5.41 Beim Lernen bin ich unkonzentriert. sehr selten ☐ ☐ ☐ ☐ sehr oft

5.42 Ich lege vor jeder Lernphase eine bestimmte Zeitdauer fest. sehr selten ☐ ☐ ☐ ☐ sehr oft

5.43 Wenn ich lerne, sorge ich dafür, dass ich in Ruhe arbeiten kann. sehr selten ☐ ☐ ☐ ☐ sehr oft

5.44 Ich stelle mir aus Mitschrift, Skript oder Literatur kurze Zusammenfassungen mit den Hauptideen zusammen. sehr selten ☐ ☐ ☐ ☐ sehr oft

5.45 Ich nehme die Hilfe anderer in Anspruch, wenn ich ernsthafte Verständnisprobleme habe. sehr selten ☐ ☐ ☐ ☐ sehr oft

5.46 Ich versuche in Gedanken, das Gelernte mit dem zu verbinden, was ich schon darüber weiß. sehr selten ☐ ☐ ☐ ☐ sehr oft

5.47 Ich lese einen Text durch und versuche, ihn mir am Ende jedes Abschnitts auswendig vorzusagen. sehr selten ☐ ☐ ☐ ☐ sehr oft

5.48 Ich stelle mir Fragen zum Stoff, um sicherzugehen, dass ich auch alles verstanden habe. sehr selten ☐ ☐ ☐ ☐ sehr oft

5.49 Gewöhnlich dauert es nicht lange, bis ich mich dazu entschließe, mit dem Lernen anzufangen. sehr selten ☐ ☐ ☐ ☐ sehr oft

5.50 Wenn ich lerne, bin ich leicht abzulenken. sehr selten ☐ ☐ ☐ ☐ sehr oft

5.51 Mein Arbeitsplatz ist so gestaltet, dass ich alles schnell finden kann. sehr selten ☐ ☐ ☐ ☐ sehr oft

Questionnaire Used in the Pre Survey 2012/2013, page 5/6

EvaSys	Fragebogen zum Vorlesungsbeginn WS 2012/13	▓ Electric Paper

5. Lernverhalten [Fortsetzung]

5.52 Ich unterstreiche in Texten oder Mitschriften die wichtigen Stellen. sehr selten ☐ ☐ ☐ ☐ sehr oft

5.53 Wenn mir etwas nicht klar ist, frage ich einen Kommilitonen um Rat. sehr selten ☐ ☐ ☐ ☐ sehr oft

5.54 Ich denke mir konkrete Beispiele zu bestimmten Lerninhalten aus. sehr selten ☐ ☐ ☐ ☐ sehr oft

5.55 Ich lerne Regeln, Fachbegriffe oder Formeln auswendig. sehr selten ☐ ☐ ☐ ☐ sehr oft

5.56 Um Wissenslücken festzustellen, rekapituliere ich die wichtigsten Inhalte, ohne meine Unterlagen zu Hilfe zu nehmen. sehr selten ☐ ☐ ☐ ☐ sehr oft

5.57 Vor der Prüfung nehme ich mir ausreichend Zeit, um den ganzen Stoff noch einmal durchzugehen. sehr selten ☐ ☐ ☐ ☐ sehr oft

5.58 Meine Konzentration hält nicht lange an. sehr selten ☐ ☐ ☐ ☐ sehr oft

5.59 Die wichtigsten Unterlagen habe ich an meinem Arbeitsplatz griffbereit. sehr selten ☐ ☐ ☐ ☐ sehr oft

5.60 Für größere Stoffmengen fertige ich eine Gliederung an, die die Struktur des Stoffs am besten wiedergibt. sehr selten ☐ ☐ ☐ ☐ sehr oft

5.61 Entdecke ich größere Lücken in meinen Aufzeichnugen, so wende ich mich an meine Kommilitonen. sehr selten ☐ ☐ ☐ ☐ sehr oft

5.62 Ich beziehe das, was ich lerne, aus meine eigenen Erfahrungen. sehr selten ☐ ☐ ☐ ☐ sehr oft

5.63 Ich lerne den Stoff anhand von Skripten oder anderen Aufzeichnungen möglichst auswendig. sehr selten ☐ ☐ ☐ ☐ sehr oft

5.64 Ich bearbeite zusätzliche Aufgaben, um festzustellen, ob ich den Stoff wirklich verstanden habe. sehr selten ☐ ☐ ☐ ☐ sehr oft

5.65 Ich nehme mir mehr Zeit zum Lernen als die meisten meiner Kommilitonen. sehr selten ☐ ☐ ☐ ☐ sehr oft

5.66 Ich stelle mir wichtige Fachausdrücke und Definitionen in eigenen Listen zusammen. sehr selten ☐ ☐ ☐ ☐ sehr oft

5.67 Ich überlege mir, ob der Lernstoff auch für mein Alltagsleben von Bedeutung ist. sehr selten ☐ ☐ ☐ ☐ sehr oft

5.68 Um mein eigenes Verständnis zu prüfen, erkläre ich bestimmte Teile des Lernstoffs einem Kommilitonen. sehr selten ☐ ☐ ☐ ☐ sehr oft

5.69 Ich arbeite so lange, bis ich mir sicher bin, die Prüfung gut bestehen zu können. sehr selten ☐ ☐ ☐ ☐ sehr oft

5.70 Wenn mir ein Teilaspekt verworren oder unklar erscheint, gehe ich ihn noch einmal langsam durch. sehr selten ☐ ☐ ☐ ☐ sehr oft

Vielen Dank für Ihre Zeit!

Questionnaire Used in the Pre Survey 2012/2013, page 6/6

EvaSys	Fragebogen zum Vorlesungsende WS 2012/2013	⊛ Electric Paper
Ruhr-Universität Bochum	Michael Kallweit	
MP² - Mathe/Plus/Praxis	MathePlus (Evaluation)	**MP**^{2}

Markieren Sie so: ☐ ☒ ☐ ☐ ☐ Bitte verwenden Sie einen Kugelschreiber oder nicht zu starken Filzstift. Dieser Fragebogen wird maschinell erfasst.

Korrektur: ☐ ■ ☐ ☒ ☐ Bitte beachten Sie im Interesse einer optimalen Datenerfassung die links gegebenen Hinweise beim Ausfüllen.

1. Informationen

Dies ist der zweite und letzte Teil der in diesem Jahr zum dritten Mal durchgeführten Befragung aller Studierenden der Vorlesungen Mathematik I für Maschinenbauer, Bauingenieure und Umwelttechniker bezüglich ihres Lernverhaltens nach dem Übergang an die Universität. Es geht in dieser Befragung vor allem darum, wie Sie nach der vergangenen Vorlesungszeit rückwirkend Ihr Arbeitsverhalten **in den Wochen nach Weihnachten** bewerten.

Im Rahmen des Projektes MP2 untersuchen wir so mit Ihrer Hilfe, welche Maßnahmen und Aspekte zu einem erfolgreichen Verlauf des ersten Studienjahrs beitragen. Die Erkenntnisse aus diesen Forschungen sollen uns helfen, Sie und zukünftige Studentinnen und Studenten effektiver zu unterstützen.

Der Zeitaufwand für den gesamten Fragebogen beträgt höchstens 15 Minuten.

Wir danken Ihnen für Ihre Kooperation!

2. Persönliche Angaben

Bitte geben Sie vor Beantwortung der Fragen den unten angegebenen Code ein.
Durch den Code wird gewährleistet, dass wir Ihre Angaben anonym mit Ihren Antworten späterer Befragungen verknüpfen können.
Alle Angaben werden nur innerhalb von MP² verwendet. Sie werden nicht an Dritte außerhalb der Projektgruppe weitergegeben, also insbesondere nicht an Personen, die Ihre Studienleistungen bewerten.

2.1 Erster Buchstabe des eigenen Vornamens (z.B. **Julia = J**)

☐ A	☐ B	☐ C
☐ D	☐ E	☐ F
☐ G	☐ H	☐ I
☐ J	☐ K	☐ L
☐ M	☐ N	☐ O
☐ P	☐ Q	☐ R
☐ S	☐ T	☐ U
☐ V	☐ W	☐ X
☐ Y	☐ Z	

2.2 Erster Buchstabe des eigenen Nachnamens (z.B. **Mustermann = M**)

☐ A	☐ B	☐ C
☐ D	☐ E	☐ F
☐ G	☐ H	☐ I
☐ J	☐ K	☐ L
☐ M	☐ N	☐ O
☐ P	☐ Q	☐ R
☐ S	☐ T	☐ U
☐ V	☐ W	☐ X
☐ Y	☐ Z	

2.3 Dritte Ziffer von rechts der Matrikelnummer (z.B. 108123456789 = 7)

☐ 0	☐ 1	☐ 2
☐ 3	☐ 4	☐ 5
☐ 6	☐ 7	☐ 8
☐ 9		

Figure A.2: Questionnaire Used in the Post Survey 2012/2013, page 1/8

EvaSys	Fragebogen zum Vorlesungsende WS 2012/2013	⬛ Electric Paper

2. Persönliche Angaben [Fortsetzung]

2.4 Zweite Ziffer von rechts der
Matrikelnummer (z.B. 108123456789 = 8)

☐ 0   ☐ 1   ☐ 2
☐ 3   ☐ 4   ☐ 5
☐ 6   ☐ 7   ☐ 8
☐ 9

2.5 Erste Ziffer von rechts der Matriklenummer
(z.B. 108123456789 = 9)

☐ 0   ☐ 1   ☐ 2
☐ 3   ☐ 4   ☐ 5
☐ 6   ☐ 7   ☐ 8
☐ 9

3. Allgemeines

Wenn Sie in diesem Semester Schwierigkeiten mit der Mathematik I Veranstaltung hatten, woran lag das Ihrer Meinung nach?

3.1 Zu große Stofffülle — stimmt genau ☐ ☐ ☐ ☐ stimmt gar nicht

3.2 Zu großer Zeitdruck — stimmt genau ☐ ☐ ☐ ☐ stimmt gar nicht

3.3 Verpflichtungen bei anderen Veranstaltungen — stimmt genau ☐ ☐ ☐ ☐ stimmt gar nicht

3.4 Vorlesung ist zu schwierig — stimmt genau ☐ ☐ ☐ ☐ stimmt gar nicht

3.5 Übungen halfen nicht weiter — stimmt genau ☐ ☐ ☐ ☐ stimmt gar nicht

3.6 Organisatorische Probleme — stimmt genau ☐ ☐ ☐ ☐ stimmt gar nicht

3.7 Meine Arbeitsweise war nicht gut genug — stimmt genau ☐ ☐ ☐ ☐ stimmt gar nicht

3.8 Sonstige Gründe:

3.9 Ich hatte keine Schwierigkeiten mit der
Mathematik I Veranstaltung.   ☐ stimmt   ☐ stimmt nicht

Wie bewerten Sie das vergangene Semester insgesamt?

3.10 Ich habe in diesem Semester meine Lernzeit effektiv genutzt — stimmt genau ☐ ☐ ☐ ☐ stimmt gar nicht

3.11 Ich habe mir genug Zeit zum Lernen genommen — stimmt genau ☐ ☐ ☐ ☐ stimmt gar nicht

3.12 Ich fühle mich gut auf die Prüfung vorbereitet — stimmt genau ☐ ☐ ☐ ☐ stimmt gar nicht

3.13 Ich bin mit meinen Arbeitstechniken zufrieden — stimmt genau ☐ ☐ ☐ ☐ stimmt gar nicht

Questionnaire Used in the Post Survey 2012/2013, page 2/8

3. Allgemeines [Fortsetzung]

3.14 Was nehmen Sie aus dem ersten Semester an Erfahrung mit?

Was wollen Sie in Zukunft besser machen?

3.15 Mehr Zeit fürs Studium nehmen — stimmt genau ☐ ☐ ☐ ☐ stimmt gar nicht

3.16 Mein Lernen stärker organisieren — stimmt genau ☐ ☐ ☐ ☐ stimmt gar nicht

3.17 Mehr mit anderen Studierenden zusammenarbeiten — stimmt genau ☐ ☐ ☐ ☐ stimmt gar nicht

3.18 Private Belastung / Arbeit reduzieren — stimmt genau ☐ ☐ ☐ ☐ stimmt gar nicht

3.19 Sonstiges:

4. Lernverhalten

Geben Sie bitte für jede im Folgenden beschriebene Tätigkeit an, wie häufig diese bei Ihnen vorgekommen ist. Sie können Ihre Antworten von **sehr selten** bis **sehr oft** abstufen.

4.1 Ich habe für mein Studium gelernt. — sehr selten ☐ ☐ ☐ ☐ sehr oft

4.2 Ich habe Tabellen, Diagramme oder Schaubilder angefertigt, um den Stoff der Veranstaltung besser strukturiert vorliegen zu haben. — sehr selten ☐ ☐ ☐ ☐ sehr oft

4.3 Ich habe versucht, Beziehungen zu den Inhalten verwandter Fächer bzw. Lehrveranstaltungen herzustellen. — sehr selten ☐ ☐ ☐ ☐ sehr oft

4.4 Ich habe mir den Lernstoff der Vorlesung durch Wiederholen eingeprägt. — sehr selten ☐ ☐ ☐ ☐ sehr oft

4.5 Ich habe versucht, mir vorher genau zu überlegen, welche Teile eines bestimmten Themengebietes ich lernen muss und welche nicht. — sehr selten ☐ ☐ ☐ ☐ sehr oft

Questionnaire Used in the Post Survey 2012/2013, page 3/8

EvaSys	Fragebogen zum Vorlesungsende WS 2012/2013	▓ Electric Paper

4. Lernverhalten [Fortsetzung]

4.6 Wenn ich schwierigen Stoff vorliegen hatte, habe ich meine Lerntechnik den höheren Anforderungen angepasst (z.B. durch langsameres Lesen). — sehr selten ☐ ☐ ☐ ☐ sehr oft

4.7 Ich habe Aufgaben zusammen mit meinen Kommilitonen bearbeitet. — sehr selten ☐ ☐ ☐ ☐ sehr oft

4.8 Ich habe nach weiterführendem Material gesucht, wenn mir bestimmte Inhalte noch nicht ganz klar waren. — sehr selten ☐ ☐ ☐ ☐ sehr oft

4.9 Wenn ich mir ein bestimmtes Pensum zum Lernen vorgenommen hatte, habe ich mich bemüht, es auch zu schaffen. — sehr selten ☐ ☐ ☐ ☐ sehr oft

4.10 Beim Lernen habe ich gemerkt, dass meine Gedanken abschweifen. — sehr selten ☐ ☐ ☐ ☐ sehr oft

4.11 Beim Lernen habe ich mich an einen bestimmten Zeitplan gehalten. — sehr selten ☐ ☐ ☐ ☐ sehr oft

4.12 Ich habe an einem Platz gelernt, wo ich mich gut auf den Stoff konzentrieren konnte. — sehr selten ☐ ☐ ☐ ☐ sehr oft

4.13 Wenn ich während des Lesens nicht alles verstanden habe, habe ich versucht, die Lücken festzuhalten und das Material daraufhin noch einmal durchzugehen. — sehr selten ☐ ☐ ☐ ☐ sehr oft

4.14 Ich habe mir kurze Zusammenfassungen der wichtigsten Inhalte als Gedankenstütze gemacht. — sehr selten ☐ ☐ ☐ ☐ sehr oft

4.15 Ich habe mir Zeit genommen, um mit Kommilitonen über den Stoff zu diskutieren. — sehr selten ☐ ☐ ☐ ☐ sehr oft

4.16 Wenn ich einen Fachbegriff nicht verstanden habe, so habe ich ihn nachgeschlagen, z.B. in einem Fachbuch oder im Internet. — sehr selten ☐ ☐ ☐ ☐ sehr oft

4.17 Zu neuen Konzepten habe ich mir praktische Anwendungen vorgestellt. — sehr selten ☐ ☐ ☐ ☐ sehr oft

4.18 Ich habe meine Aufzeichnungen / das Skript mehrmals hintereinander durchgelesen. — sehr selten ☐ ☐ ☐ ☐ sehr oft

4.19 Ich habe im Vorhinein festgelegt, wie weit ich mit der Durcharbeitung des Stoffes kommen wollte. — sehr selten ☐ ☐ ☐ ☐ sehr oft

4.20 Ich habe mich auch dann angestrengt, wenn mir der Stoff überhaupt nicht lag. — sehr selten ☐ ☐ ☐ ☐ sehr oft

4.21 Es ist mir schwergefallen, bei der Sache zu bleiben. — sehr selten ☐ ☐ ☐ ☐ sehr oft

Questionnaire Used in the Post Survey 2012/2013, page 4/8

4. Lernverhalten [Fortsetzung]

4.22 Ich habe bestimmte Zeiten festgelegt, zu denen ich dann gelernt habe.
sehr selten ☐ ☐ ☐ ☐ sehr oft

4.23 Ich habe meine Umgebung so gestaltet, dass ich möglichst wenig vom Lernen abgelenkt wurde.
sehr selten ☐ ☐ ☐ ☐ sehr oft

4.24 Ich bin meine Aufzeichnungen durchgegangen und habe mir dazu eine Gliederung mit den wichtigsten Punkten gemacht.
sehr selten ☐ ☐ ☐ ☐ sehr oft

4.25 Ich habe meine Vorlesungsmitschriften mit denen meiner Kommilitonen verglichen.
sehr selten ☐ ☐ ☐ ☐ sehr oft

4.26 Fehlende Informationen habe ich mir aus verschiedenen Quellen zusammengesucht (z.B. Internet, Bücher, Fachzeitschriften).
sehr selten ☐ ☐ ☐ ☐ sehr oft

4.27 Ich habe versucht, neue Begriffe oder Theorien auf mir bereits bekannte Begriffe und Theorien zu beziehen.
sehr selten ☐ ☐ ☐ ☐ sehr oft

4.28 Ich habe Schlüsselbegriffe auswendiggelernt, um mich in der Prüfung besser an wichtige Inhaltsbereiche erinnern zu können.
sehr selten ☐ ☐ ☐ ☐ sehr oft

4.29 Vor dem Lernen eines Stoffgebietes habe ich mir überlegt, wie ich am effektivsten vorgehen kann.
sehr selten ☐ ☐ ☐ ☐ sehr oft

4.30 Ich habe nicht aufgegeben, auch wenn der Stoff sehr schwierig oder komplex wurde.
sehr selten ☐ ☐ ☐ ☐ sehr oft

4.31 Ich habe mich dabei ertappt, dass ich mit meinen Gedanken ganz woanders war.
sehr selten ☐ ☐ ☐ ☐ sehr oft

4.32 Ich habe die Stunden, die ich täglich mit Lernen verbringen wollte, durch einen Zeitplan festgelegt.
sehr selten ☐ ☐ ☐ ☐ sehr oft

4.33 Zum Lernen habe ich immer am selben Platz gesessen.
sehr selten ☐ ☐ ☐ ☐ sehr oft

4.34 Ich habe versucht, den Stoff so zu ordnen, dass ich ihn mir gut einprägen konnte.
sehr selten ☐ ☐ ☐ ☐ sehr oft

4.35 Ich habe mich von einem Kommilitonen abfragen lassen und habe auch ihm Fragen zum Stoff gestellt.
sehr selten ☐ ☐ ☐ ☐ sehr oft

4.36 Ich habe zusätzliche Quellen herangezogen, wenn meine Aufzeichnungen unvollständig waren.
sehr selten ☐ ☐ ☐ ☐ sehr oft

4.37 Ich habe mir manche Sachverhalte bildlich vorgestellt.
sehr selten ☐ ☐ ☐ ☐ sehr oft

Questionnaire Used in the Post Survey 2012/2013, page 5/8

EvaSys	Fragebogen zum Vorlesungsende WS 2012/2013	▓ Electric Paper

4. Lernverhalten   [Fortsetzung]

4.38 Ich habe eine selbst erstellte Übersicht mit den wichtigsten Fachtermini auswendiggelernt.   sehr selten ☐ ☐ ☐ ☐ sehr oft

4.39 Ich habe mir vorher überlegt, in welcher Reihenfolge ich den Stoff durcharbeiten wollte.   sehr selten ☐ ☐ ☐ ☐ sehr oft

4.40 Ich habe auch spätabends und am Wochenende gelernt, wenn es sein musste.   sehr selten ☐ ☐ ☐ ☐ sehr oft

4.41 Beim Lernen war ich unkonzentriert.   sehr selten ☐ ☐ ☐ ☐ sehr oft

4.42 Ich habe vor jeder Lernphase eine bestimmte Zeitdauer festgelegt.   sehr selten ☐ ☐ ☐ ☐ sehr oft

4.43 Wenn ich gelernt habe, habe ich dafür gesorgt, dass ich in Ruhe arbeiten konnte.   sehr selten ☐ ☐ ☐ ☐ sehr oft

4.44 Ich habe mir aus Mitschrift, Skript oder Literatur kurze Zusammenfassungen mit den Hauptideen zusammengestellt.   sehr selten ☐ ☐ ☐ ☐ sehr oft

4.45 Ich habe die Hilfe anderer in Anspruch genommen, wenn ich ernsthafte Verständnisprobleme hatte.   sehr selten ☐ ☐ ☐ ☐ sehr oft

4.46 Ich habe in Gedanken versucht, das Gelernte mit dem zu verbinden, was ich schon darüber wusste.   sehr selten ☐ ☐ ☐ ☐ sehr oft

4.47 Ich habe einen Text durchgelesen und versucht, ihn mir am Ende jedes Abschnitts auswendig vorzusagen.   sehr selten ☐ ☐ ☐ ☐ sehr oft

4.48 Ich habe mir Fragen zum Stoff gestellt, um sicherzugehen, dass ich auch alles verstanden habe.   sehr selten ☐ ☐ ☐ ☐ sehr oft

4.49 Gewöhnlich hat es nicht lange gedauert, bis ich mich dazu entschlossen habe, mit dem Lernen anzufangen.   sehr selten ☐ ☐ ☐ ☐ sehr oft

4.50 Wenn ich gelernt habe, war ich leicht abzulenken.   sehr selten ☐ ☐ ☐ ☐ sehr oft

4.51 Mein Arbeitsplatz war so gestaltet, dass ich alles schnell finden konnte.   sehr selten ☐ ☐ ☐ ☐ sehr oft

4.52 Ich habe in Texten oder Mitschriften die wichtigen Stellen unterstrichen.   sehr selten ☐ ☐ ☐ ☐ sehr oft

4.53 Wenn mir etwas nicht klar war, habe ich einen Kommilitonen um Rat gefragt.   sehr selten ☐ ☐ ☐ ☐ sehr oft

4.54 Ich habe mir konkrete Beispiele zu bestimmten Lerninhalten ausgedacht.   sehr selten ☐ ☐ ☐ ☐ sehr oft

---

Questionnaire Used in the Post Survey 2012/2013, page 6/8

4. Lernverhalten   [Fortsetzung]

4.55 Ich habe Regeln, Fachbegriffe oder Formeln auswendiggelernt.   sehr selten ☐ ☐ ☐ ☐ sehr oft

4.56 Um Wissenslücken festzustellen, habe ich die wichtigsten Inhalte, ohne meine Unterlagen zu Hilfe zu nehmen, rekapituliert.   sehr selten ☐ ☐ ☐ ☐ sehr oft

4.57 Vor der Klausur habe ich mir ausreichend Zeit eingeplant, um den ganzen Stoff noch einmal durchzugehen.   sehr selten ☐ ☐ ☐ ☐ sehr oft

4.58 Meine Konzentration hat nicht lange angehalten.   sehr selten ☐ ☐ ☐ ☐ sehr oft

4.59 Die wichtigsten Unterlagen habe ich an meinem Arbeitsplatz griffbereit gehabt.   sehr selten ☐ ☐ ☐ ☐ sehr oft

4.60 Für größere Stoffmengen habe ich eine Gliederung angefertigt, die die Struktur des Stoffs am besten wiedergibt.   sehr selten ☐ ☐ ☐ ☐ sehr oft

4.61 Habe ich größere Lücken in meinen Aufzeichnungen entdeckt, so habe ich mich an meine Kommilitonen gewandt.   sehr selten ☐ ☐ ☐ ☐ sehr oft

4.62 Ich habe das, was ich lernte, auf meine eigenen Erfahrungen bezogen.   sehr selten ☐ ☐ ☐ ☐ sehr oft

4.63 Ich habe den Stoff anhand von Skripten oder anderen Aufzeichnungen möglichst auswendiggelernt.   sehr selten ☐ ☐ ☐ ☐ sehr oft

4.64 Ich habe zusätzliche Aufgaben bearbeitet, um festzustellen, ob ich den Stoff wirklich verstanden habe.   sehr selten ☐ ☐ ☐ ☐ sehr oft

4.65 Ich habe mir mehr Zeit zum Lernen genommen als die meisten meiner Kommilitonen.   sehr selten ☐ ☐ ☐ ☐ sehr oft

4.66 Ich habe mir wichtige Fachausdrücke und Definitionen in eigenen Listen zusammengestellt.   sehr selten ☐ ☐ ☐ ☐ sehr oft

4.67 Ich habe mir überlegt, ob der Lernstoff auch für mein Alltagsleben von Bedeutung ist.   sehr selten ☐ ☐ ☐ ☐ sehr oft

4.68 Um mein eigenes Verständnis zu prüfen, habe ich bestimmte Teile des Lernstoffs einem Kommilitonen erklärt.   sehr selten ☐ ☐ ☐ ☐ sehr oft

4.69 Ich habe so lange gearbeitet, bis ich mir sicher war, eine Klausuraufgabe zu diesem Thema gut schaffen zu können.   sehr selten ☐ ☐ ☐ ☐ sehr oft

Questionnaire Used in the Post Survey 2012/2013, page 7/8

| EvaSys | Fragebogen zum Vorlesungsende WS 2012/2013 | ● Electric Paper |

4. Lernverhalten [Fortsetzung]

4.70 Wenn mir ein Teilaspekt verworren oder unklar erschien, bin ich ihn noch einmal langsam durchgegangen.

sehr selten ☐ ☐ ☐ ☐ sehr oft

5. Ende

Vielen Dank und viel Erfolg bei den in den nächsten Wochen anstehenden Prüfungen wünscht Ihnen das MP²-Team!

Questionnaire Used in the Post Survey 2012/2013, page 8/8

B LIST Questionnaire, English

Here, the items from the LIST questionnaire, as used in the pre and in the post survey, are ordered by scale (translation by author). In the surveys themselves, they were mixed, and supplemented by other questions, which were different in the pre and in the post surveys.

Pre Survey

1. Organizing

   a) I make charts, diagrams and graphics in order to have the subject matter in front of me in a structured form.

   b) I compile short summaries of the most important contents as a mnemonic aid.

   c) I go over my notes and structure the most important points.

   d) I try to order the subject matter in a way that makes it easy for me to remember.

   e) I compile a summary of the main ideas out of my notes, the script or other sources.

   f) I underline the most important parts in my notes or in the texts.

   g) For bigger amounts of subject matter I find an arrangement that mirrors the structure best.

   h) I assemble important terms and definitions in my own lists.

2. Elaborating

   a) I try to find connections to other subjects or courses.

   b) I think of practical applications of new concepts.

   c) I try to relate new terms or theories to terms or theories I already know.

   d) I visualize new issues.

   e) In my mind I try to connect newly learnt facts to what I already know.

   f) I think of practical examples for certain curricular facts.

g) I relate what I am learning to my own experiences.

h) I wonder if the subject matter is relevant for my everyday life.

3. Repeating

a) I imprint the subject matter from the lecture on my memory by repeating it.

b) I read my notes several times in a row.

c) I learn key terms by heart in order to remember important facts better in the exam.

d) I commit a self-compiled compendium to memory.

e) I read a text and try to recite it at the end of each paragraph.

f) I commit rules, technical terms or formulas to memory.

g) I learn the subject matter by heart using scripts or other notes.

4. Metacognition: Planning

a) I try to consider beforehand which areas of certain topics I have to study and which I do not have to study.

b) I decide in advance how much subject matter I would like to work through in this session.

c) Before starting on an area of expertise, I reflect upon how to work most efficiently.

d) I plan in advance in which order I want to work through the subject matter.

5. Metacognition: Monitoring

a) I ask myself questions on the subject matter in order to make sure that I have understood everything correctly.

b) In order to find gaps in my knowledge I sum up the most important contents without using my notes.

c) I work on additional tasks in order to determine if I have truly understood the subject matter.

d) In order to check my own understanding I explain certain parts of the subject matter to a fellow student.

6. Metacognition: Regulating

    a) If I am confronted with a difficult subject matter, I will adapt my learning technique to the higher demands. Confronted with a difficult subject matter I adapt my learning strategy accordingly.

    b) If I do not understand everything I am reading, I will try to make a note of the gap in my knowledge and sift through the material again.

    c) When an aspect seems confusing or unclear, I examine it again thoroughly.

7. Effort

    a) Whenever I have planned a certain workload, I make an effort to master it.

    b) I make an effort even though the subject matter may not suit me well.

    c) I do not give up even though the subject matter is very difficult and complex.

    d) I work late at night or at the weekends if necessary.

    e) It usually does not need much time until I decide to start working.

    f) Before exams I take the time to go over all the subject matter again.

    g) I take more time for learning than most of my fellow students.

    h) I work until I am sure to pass the exam well.

8. Attention

    a) When I am learning I notice that my thoughts tend to stray.

    b) It is difficult for me to concentrate.

    c) I find myself thinking of completely different things.

    d) When learning I am lacking in concentration.

    e) I am easy to distract when learning.

    f) My concentration does not last very long.

9. Time management

    a) I work according to a schedule.

b) I decide on the times for my learning.

c) I fix the hours I spend daily on learning in a schedule.

d) Before each study period I appoint the duration of my work.

10. Learning environment

a) I work in a place that makes it easy to concentrate.

b) I design my work environment in a way that I am distracted as little as possible.

c) When learning I always sit at the same place.

d) When studying I make sure that I can work uninterrupted.

e) My workplace is designed in a way that makes it easy to find everything.

f) At my desk I have the most important papers within reach.

11. Peer learning

a) I work on tasks together with my fellow students.

b) I take my time to discuss the subject matter with my fellow students.

c) I compare my notes with my fellow students'.

d) I make fellow students ask me questions on the subject matter and ask them questions too.

e) I turn to help from others when I have serious problems in understanding something.

f) When I am not sure about something I ask a fellow student for advice.

g) If I find considerable gaps in my notes, I turn to fellow students.

12. Using works of reference

a) I search for explanatory material if certain facts are not completely clear.

b) Whenever I do not understand a technical term, I look it up in a textbook or on the Internet.

c) I look for missing information in different sources, e.g. the Internet, textbooks, or journals.

d) When my notes are incomplete I use additional sources.

Post Survey

1. Organizing
   a) I made charts, diagrams and graphics in order to have the subject matter in front of me in a structured form.
   b) I compiled short summaries of the most important contents as a mnemonic aid.
   c) I went over my notes and structured the most important points.
   d) I tried to order the subject matter in a way that made it easy for me to remember.
   e) I compiled a summary of the main ideas out of my notes, the script or other sources.
   f) I underlined the most important parts in my notes or in the texts.
   g) For bigger amounts of subject matter I found an arrangement that mirrors the structure best.
   h) I assembled important terms and definitions in my own lists.

2. Elaborating
   a) I tried to find connections to other subjects or courses.
   b) I thought of practical applications of new concepts.
   c) I tried to relate new terms or theories to terms or theories I already knew.
   d) I visualized new issues.
   e) In my mind I tried to connect newly learnt facts to what I already knew.
   f) I thought of practical examples for certain curricular facts.
   g) I related what I was learning to my own experiences.
   h) I wondered if the subject matter was relevant for my everyday life.

3. Repeating
   a) I imprinted the subject matter from the lecture on my memory by repeating it.

b) I read my notes several times in a row.

c) I learned key terms by heart in order to remember important facts better in the exam.

d) I committed a self-compiled compendium to memory.

e) I read a text and tried to recite it at the end of each paragraph.

f) I committed rules, technical terms or formulas to memory.

g) I learned the subject matter by heart using scripts or other notes.

4. Metacognition: Planning

a) I tried to consider beforehand which areas of certain topics I had to study and which I did not have to study.

b) I decided in advance how much subject matter I would like to work through in each session.

c) Before starting on an area of expertise, I reflected upon how to work most efficiently.

d) I planned in advance in which order I wanted to work through the subject matter.

5. Metacognition: Monitoring

a) I asked myself questions on the subject matter in order to make sure that I had understood everything correctly.

b) In order to find gaps in my knowledge I summed up the most important contents without using my notes.

c) I worked on additional tasks in order to determine if I had truly understood the subject matter.

d) In order to check my own understanding I explained certain parts of the subject matter to a fellow student.

6. Metacognition: Regulating

a) If I was confronted with a difficult subject matter, I would adapt my learning technique to the higher demands. Confronted with a difficult subject matter I adapted my learning strategy accordingly.

b) If I did not understand everything I was reading, I would try to make a note of the gap in my knowledge and sift through the material again.

c) When an aspect seemed confusing or unclear, I examined it again thoroughly.

7. Effort

a) Whenever I had planned a certain workload, I made an effort to master it.

b) I made an effort even though the subject matter did not suit me well.

c) I did not give up even though the subject matter was very difficult and complex.

d) I worked late at night or at the weekends when necessary.

e) It usually did not need much time until I decided to start working.

f) Before exams I took the time to go over all the subject matter again.

g) I took more time for learning than most of my fellow students.

h) I worked until I was sure to pass the exam well.

8. Attention

a) When I was learning I noticed that my thoughts tended to stray.

b) It was difficult for me to concentrate.

c) I found myself thinking of completely different things.

d) When learning I was lacking in concentration.

e) I was easy to distract when learning.

f) My concentration did not last very long.

9. Time management

a) I worked according to a schedule.

b) I decided on the times for my learning.

c) I fixed the hours I spent daily on learning in a schedule.

d) Before each study period I appointed the duration of my work.

10. Learning environment

   a) I worked in a place that made it easy to concentrate.

   b) I designed my work environment in a way that I was distracted as little as possible.

   c) When learning I always sat at the same place.

   d) When studying I made sure that I could work uninterrupted.

   e) My workplace was designed in a way that made it easy to find everything.

   f) At my desk I had the most important papers within reach.

11. Peer learning

   a) I worked on tasks together with my fellow students.

   b) I took my time to discuss the subject matter with my fellow students.

   c) I compared my notes with my fellow students'.

   d) I made fellow students ask me questions on the subject matter and asked them questions too.

   e) I turned to help from others when I had serious problems in understanding something.

   f) When I was not sure about something I asked a fellow student for advice.

   g) If I found considerable gaps in my notes, I turned to fellow students.

12. Using works of reference

   a) I searched for explanatory material if certain facts were not completely clear.

   b) Whenever I did not understand a technical term, I looked it up in a textbook or on the Internet.

   c) I looked for missing information in different sources, e.g. the Internet, textbooks, or journals.

   d) When my notes were incomplete I used additional sources.

C Evaluation Survey 2010/2011

The evaluation survey was conducted after all project interventions (except the revision course) had been completed, before the examination. The questions start with the anonymous participation code, consisting of

- first letter of mother's given name,
- day of mother's birthday in two digits,
- first letter of father's given name,
- day of father's birthday in two digits,
- first letter of student's given name, and
- day of student's birthday in two digits.[1]

This was meant to enable connections between different surveys and also link the surveys anonymously to results in the examination. Unfortunately, this participation code proved not practicable, as it resulted far too many different codes (across the different surveys, including pre and post as well as the evaluation survey and the learning diaries) than actually possible for the number of students enrolled.

The questions were grouped according to the intervention they referred to. The answer options ranged from 4-point-Likert scales to yes / no and free text.

1. Preparatory tutorials
   a) How did you like the size of groups?
   b) How do you rate the atmosphere during group sessions?
   c) How do you rate the subject-specific support?
   d) How do you rate the methodical support?
   e) How do you rate the general concept of the preparatory tutorials?
   f) Did you get help in the preparatory tutorials?

---

[1] The example given in the survey was Julia Mustermann, born July 25th, with mother Else, born November 5th, and father Klaus, born on January 14th, resulting in the participation code E05K14J25.

g) What, in your opinion, could be improved?

h) Would you recommend the preparatory tutorials?

i) Would you like to tell us something else about the preparatory tutorials?

2. MP^2-HelpDesk

a) Did you go to the MP^2-Math/Plus HelpDesk?

b) If so, how often?

c) How content were you with the opening times (frequency, times)?

d) Did you get help at the MP^2-HelpDesk?

e) What, in your opinion, could be improved?

f) Would you recommend the MP^2-HelpDesk?

g) Would you like to tell us something else about the MP^2-HelpDesk?

3. SZMA HelpDesk

a) Did you go to the SZMA HelpDesk?

b) If so, how often?

c) How content were you with the opening times (frequency, times)?

d) Did you get help at the SZMA-HelpDesk?

e) What, in your opinion, could be improved?

f) Would you recommend the SZMA-HelpDesk?

g) Would you like to tell us something else about the MP^2-HelpDesk?

4. e-Learning course

a) How often did you log into the e-learning course?

b) How often did you work on tasks (tests) there?

c) Were the tasks helpful?

d) How often did you look up other material?

e) Was this other material helpful?

f) Did you upload material?

g) What, in your opinion, could be improved?

h) Would you recommend the e-learning course?

i) Would you like to tell us something else about the MP^2-HelpDesk?

5. Learning diary

a) Did the learning diary help you to structure your learning?

b) Hoe do you rate the learning diary in reference to how often it had to be filled in?

c) How do you rate the learning diary in reference to quantity?

d) What, in your opinion, could be improved?

e) Would you recommend the learning diary?

f) Would you like to tell us something else about the learning diary?

6. MP^2 in general

a) Did participating in MP^2-Math/Plus help you?

b) What, in your opinion, could be improved?

c) Would you recommend MP^2-Math/Plus?

d) Would you like to tell us something else about MP^2 ?

The survey closes with thanks for filling in the form and best wishes for the examination.

D Learning Diary 2010/2011

The learning diary was meant to be filled in every day, including weekends and holidays. Students were allowed to choose between an online and a paper version. Like most surveys, it was compiled in EvaSys[2], a survey and evaluation software. After some initial information, the participation code was required, see appendix C. To arrange the data later in subsequent order, the date was asked. The following questions included 5-point Likert scales, single choice, and a filter question (Do you intend to learn for mathematics today?) whose answer led to different blocks of other questions.

1. Part 1: Before learning

    How appropriate are the following statements for you today?

    a) Today I am very busy because of jobs I do beside university.

    b) Today I am very busy because of the demands of my university course.

    c) Today I am very busy because of private projects.

    d) Today I particularly want to relax in my free time.

    e) How do you rate your motivation to learn for mathematics today? (11-point scale from 0 to 100)

    f) Do you intend to learn for mathematics today? (If not, please continue with the questions in the last part. If so, please continue with the questions in the following part.)

2. What aims do you want to reach today when learning mathematics?

    a) Will you learn on your own today or with others?

    How appropriate are the following statements for you today?

    b) Today I particularly want to achieve a great deal.

    c) It is important that I reach my learning aims today.

    d) I am sure to reach my learning aims today.

---

[2] https://www.evasys.de

e) What do want to do in order to reach your aims today? (e.g. make an extra effort, plan my time beforehand, think of partial successes and rewards for them, avoid being disrupted)

I am motivated to learn for mathematics today because ...

f) I feel like it.

g) it is important to me today to pass the examination.

h) I know exactly what to do.

i) learning for mathematics worked out well last time.

j) I feel good today.

k) I am not busy with other thing today.

l) I planned to learn for mathematics today.

How appropriate are the following statements for you today?

m) Today, before learning, I considered how to proceed most efficiently.

n) While learning I will remind myself of my aims again and again.

o) While learning today, I will take care if I can reach my learning aim with my learning behaviour.

p) When confronted with a difficult task today, I will cope with it.

q) I will not give up while learning today, no matter if the task is difficult and does not suit me well.

r) Today I will finish what I planned to do.

3. Part 2: After learning

a) Where did you learn today? (options: on campus, at home, at a fellow student's place, somewhere else)

How appropriate are the following statements for you today?

b) I had problems to get started with learning today.

c) I had problems to keep up my motivation for learning today.

d) Today I lost sight of my learning goals.

e) I made an extra effort at learning today.

f) Today I lacked concentration while learning.

g) Today I was able to prohibit irrelevant things.

h) Today I made sure I could learn quietly.

i) Today I was able to adapt my learning style to the demands of the tasks (e.g. by adapting my pace of work).

j) Today I postponed some of what I had planned to tomorrow or later.

What is the result of your learning today?

k) What percentage of your workload planned for today did you accomplish?

l) How much time (in hours) did you spend on learning today altogether?

m) How much time (in hours) of that did you spend learning effectively? (Effective learning time refers to the time you really concentrated to the point, no matter if you are satisfied with the result or not.)

How do you rate you learning success today?

n) I am content with the relation between effective learning time and learning time altogether.

o) I overestimated my capacity today.

p) I worked qualitatively well today.

q) Today I reached my personal learning goals set before learning.

r) In your estimation, which measures or modes of behaviour were instrumental for reaching your learning goals? (e.g. setting realistic goals, planning to reward myself)

s) To your mind, what hindered you reaching your learning goals? (e.g. planning to do too much, being disrupted, being unaware of the difficulties)

What do you plan to do differently when learning next?

t) I will set myself lower learning goals.

u) I will set myself higher learning goals.

v) I intend to optimise my learning behaviour.

w) What do plan to do specifically in order to reach your learning aims? (e.g. relocate my workplace, make regular breaks, organise help)

4. I am not motivated to learn mathematics because ...

    a) I don't feel like it.

    b) it is not important to me today to pass the examination.

    c) I don't know what to do.

    d) learning for mathematics didn't worked out at all last time.

    e) I feel bad today.

    f) private matters keep me busy today.

    g) I want to devote my time today to other subjects.

    h) I haven't planned to learn for any subject today.

E Support Agreement from the Second Project Cycle

Supported Learning Group

Between _____ (participant) and the project MP^2 , represented by _____ (project manager / project tutor), this support agreement is concluded in reference to Math/Plus in the winter semester 2011/2012.

This agreement contains mutual obligations. In detail, the following issues are agreed:

For the participant:

1. The participant is obliged to consistently attend the regular sessions (lectures and tutorials) of *Mathematics for MB, BI/UTRM* and actively take part in them. This also covers handing in the weekly assignments.

2. The participant is obliged to consistently attend the additional sessions of MP^2-Math/Plus and actively take part in them. This covers meetings of the learning group, visiting the MP^2-Math/Plus helpdesk, and logging into the e-learning course.

3. In case the participant is unable to meet these obligations, he / she will inform the project manager of his / her reasons of their own accord.

4. The participant is obliged to fill in the weekly workbooks and the questionnaires provided within a given time.

In case of repeated violation against single issues of this agreement, expulsion from the project can be effected.

For the project staff:

1. The project staff are obliged to be regularly available for the participant for problems and requests concerning the above-mentioned events.

2. The project staff and management will provide qualified help in the learning group sessions, competent support in the MP^2-Math/Plus helpdesk, and expert maintenance of the e-learning course.

3. The project staff and management is obliged to respond to the participant's suggestions at all times.

4. The project management are obliged to only process the personal data
collected for evaluation and scientific purposes, and later retain them in
anonymous form.

F Questions to the Mentors

The following file was mailed to the mentors before compiling the weekly workbooks, in alle project cycles but the very first.

Dear mentors!

Thank you very much for your willingness to lend your support in words and deeds to this cycle's first-years. We would like to ask you for your statements for the workbooks. Answering is quick and easy: just write your answer under the question. If you cannot think of anything, that is fine, just skip that particular question. Please answer individually, so we can match each answer to a person. Thanks!

1. How do you remember MP^2?

2. How did you motivate yourself to carry on during the semester?

3. How did you motivate yourself to carry on learning for the examination?

4. How did you plan your weeks during the semester?

5. What comes to your mind when you think of complex numbers?

6. What did you like best in the first semester?

7. Which learning rituals did you establish?

8. What mathematics book can you recommend?

9. What did you use for secondary literature?

10. Where do you like learning best?

11. How did you rework the lecture?

12. How much time did you invest in mathematics?

13. How do you start a learning session?

14. How do you reward yourself when you have reached a goal?

15. How did you take notes?

16. How did you work on the weekly assignments?

17. How did you learn? Did you use flashcards?

18. How did you visualise the subject matter? Mindmaps?

19. What was the most difficult topic for you in mathematics 1?

20. Where did you fetch help?

21. What resources did you use for mathematics?

22. Why did you want to study engineering?

23. What do you think, what helped most? What helped you to pass the examination?

24. What should everyone master in any case (concerning mathematics 1)?

25. What tasks always go wrong for you?

26. What was your learning team like?

27. How much time did you spend on mathematics every week?

28. How much time should you have spent on mathematics every week?

29. How do you recover from learning stress?

30. What advice can you give to first-years?

31. What did you like about mathematics?

32. What tasks worked best for you?

33. Which MP^2-Math/Plus intervention helped you most?

34. Who helped you most?

35. A few words on the lecture.

36. A few words on the tutorials.

37. A few words on the weekly assignments.

38. A few words on the examination.

39. A few words on your fellow students.

40. How did you like your first university mathematics examination?

41. How long did it take you to mentally arrive at university?

42. From when on did uni become routine for you?

43. How many tasks did you solve in preparation for the examination?

44. What does you work environment look like?

G Workbook Topics

This appendix gives an overview of the *Workbooks* introduced in the second project cycle.

Start-up Aid (*Starthilfe*) With the help of colour photos, the *Mentors* are introduced, reminiscing about MP^2. MP^2-Math/Plus is explained in detail and the various project interventions and activities are presented, including information about the MP^2-Math/Plus *HelpDesk* times and rooms. The rules for participating in the project are summarised. All MP^2-Math/Plus contacts are listed, with photos and e-mail addresses. A collection of statements by the *Mentors* forms one of the recurring categories that are repeated in all subsequent *Workbooks*. The *Workbook* is completed by space to fill in for university contacts, a preview for the next week and the wishing well / suggestion box. Another element is the seven truths according to Zucker (1996), such as "You are no longer in high school", "Expect to have material covered at *two to three* times the pace of high school", or "It is *your* responsibility to learn the material" (p. 865, emphases from the original).

Navigation Aid (*Navigationshilfe*) The concept of the MP^2-Math/Plus *HelpDesk* in contrast to the SZMA helpdesk is explained, as well as the e-learning course and the learning log. MP^2-Math/Plus' presence in social networks is presented. First concrete tips for motivation and perseverance are given by suggesting setting small specific aims, planning in written form, discussing mathematics and working regularly. To specify this, the techniques *Don't break the chain* (http://dontbreakthechain.com) and *Pomodoro* (http://www.pomodorotechnique.com) are recommended, together with a graphic suggestion of how to divide aims into manageable small portions. What is more, the students are encouraged to look back on the mathematics lecture so far, collect relevant keywords and rank them according to importance. This results in a list giving the information on what they (individually) already know and what they have to work on in more detail.

Upshift (*Hochgeschaltet*) This *Workbook* concentrates on tips for time management, including a weekly and a semester plan. It also collects ideas for finding access to learning, suitable places for learning and reminded students of setting themselves specific learning targets as motivational aid.

Training Victory (*Trainingssieg*) Here the focus is on note-taking. A system for taking structured notes is introduced, together with tips like using sticky notes, abbreviations, table of contents, keywords or colors. For the omnipresent need for learning facts by heart, basics on learner types are given and learning with flashcards are recommended.

Pole Position This *Workbook* falls into the time directly before the Christmas break and marks the end of the first phase which means that the students were to cope more on their own. Both reasons find their expression in the fact that *Pole Position* concentrates on how to cope during non-term time when there are no meetings in the *Learning Group* and no lectures or tutorials. The remedy is to get digital help from the computer in the form of Computer Algebra Systems (CAS), GeoGebra, or WolframAlpha. Some examples how to do so from the homework exercises are demonstrated in detail.

Pit Stop (*Boxenstopp*) To get an overview on what lies behind and what is still to come, this *Workbook* contains an intermediate conclusion, so that students can create an overview over the subject matter (basic formulae and rules, divided into the main parts linear algebra and analysis), in combination with a structure of how well they had mastered certain areas.

Cruise Control (*Tempomat*) Apart from some information on MP^2-Math/Plus activities during non-term time (dates for *Revision Course*, the *HelpDesk*, the *Learning Group*), this *Workbook* concentrates on work on old examination papers. Before simply solving all exercises, *Cruise Control* inspires making a list counting which type of exercise occurred how often, thus producing a priority list that can be used for the examination preparations.

Home Stretch (*Zielgerade*) As the written examination is drawing near, this time the focus is on getting mathematics on paper, i.e. tips for writing down solutions and avoiding mistakes. There is also help to identify one's own typical mistakes in order to avoid them in future.

Final Spurt The very last *Workbook* (though not the last meeting of the *Learning Group*) aims even more at preparing for the written test. It contains a to-do list and a weekly timetable till the written test, together with tips for a successful test policy and an interview with Nadine (a teaching assistant) telling about typical mistakes to be avoided. *Final Spurt* ends with a motivating list to be filled in of what treats might serve as a reward after the test.

H Further Tables for Examination Statistics

The Tables H.1, H.2, and H.3 , respectively the Figures H.1, H.2, and H.3 verify that the tendency that females profit more from MP^2-Math/Plus than males cannot be detected in the fourth project cycle from 2013/2014, see section 5.5.

Table H.4 and Figure H.4 are presented to ensure comparability with previous project cycles, when the courses ET/IT and ITS were not involved in MP^2-Math/Plus.

Table H.1: Examination Statistics for the Fourth Project Cycle in 2013/2014, MP^2-Math/Plus Participants Opposed to Non-Participants, Only BI/UTRM

	pass	fail	sum
MP^2 participants	67 (=35+32)	20 (=7+13)	87 (=42+45)
non-participants	224 (=132+92)	73 (=43+30)	297 (=175+122)
sum	291 (=167+124)	93 (=50+43)	384 (=217+167)

The numbers in brackets are for males and females only, respectively.

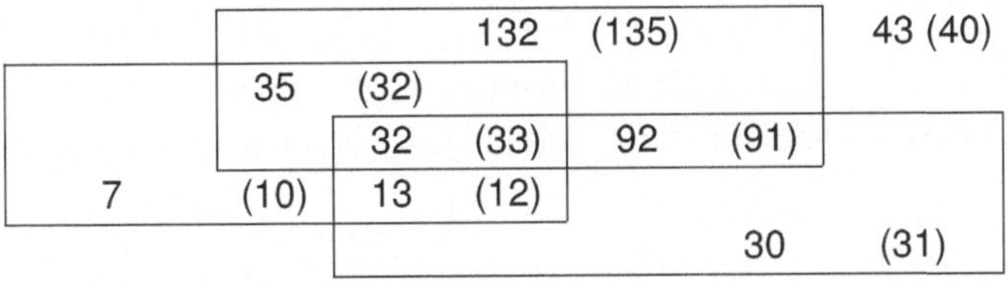

Figure H.1: Venn Diagram for Examination Statistics for the Fourth Project Cycle in 2013/2014, MP^2 Participants Opposed to Non-Participants (top centre set: passed exam, bottom left set: MP^2 participant, bottom right set: female), only BI/UTRM

Table H.2: Examination Statistics for the Fourth Project Cycle in 2013/2014,
MP^2 Participants Opposed to Non-Participants, only MB

	pass	fail	sum
MP^2 participants	11 (=3+8)	4 (=2+2)	15 (=5+10)
non-participants	265 (=242+23)	85 (=71+14)	350 (=313+37)
sum	276 (=245+31)	89 (=73+16)	365 (=318+47)

The numbers in brackets are for males and females only, respectively.

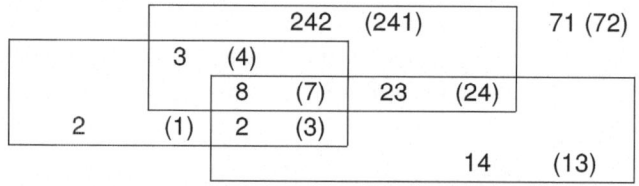

Figure H.2: Venn Diagram for Examination Statistics for the Fourth Project
Cycle in 2013/2014, MP^2 Participants Opposed to Non-Participants
(top centre set: passed exam, bottom left set: MP^2 participant,
bottom right set: female), only MB

Table H.3: Examination Statistics for the Fourth Project Cycle in 2013/2014,
MP^2 Participants Opposed to Non-Participants, only ET/IT and ITS

	pass	fail	sum
MP^2 participants	17 (=16+1)	15 (=12+3)	32 (=28+4)
non-participants	95 (=83+12)	150 (=140+10)	245 (=223+22)
sum	112 (=99+13)	165 (=152+13)	277 (=251+26)

The numbers in brackets are for males and females only, respectively.

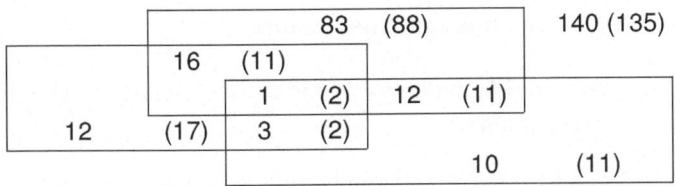

Figure H.3: Venn Diagram for Examination Statistics 2013/2014, MP^2 Participants Opposed to Non-Participants (top centre set: passed exam, bottom left set: MP^2 participant, bottom right set: female), only ET/IT, ITS

Table H.4: Examination Statistics for the Fourth Project Cycle in 2013/2014, MP^2 Participants Opposed to Non-Participants, ET/IT and ITS excluded

	pass	fail	sum
MP^2 participants	78 (=38+40)	24 (=9+15)	102 (=47+55)
non-participants	489 (=374+115)	158 (=114+44)	647 (=488+159)
sum	567 (=412+155)	182 (=123+59)	749 (=35+214)

The numbers in brackets are for males and females only, respectively.

```
                    374    (376)        114 (112)
            38   (36)
                    40    (40)   115   (115)
     9     (11)   15    (15)
                                  44        (44)
```

Figure H.4: Venn Diagram for Examination Statistics for the Fourth Project Cycle in 2013/2014, MP^2 Participants Opposed to Non-Participants (top centre set: passed exam, bottom left set: MP^2 participant, bottom right set: female), ET/IT and ITS excluded

I LIST Items and Loadings for Nine Factors

As depicted in section 5.6, the items and their loadings on the nine remaining factors are shown in Table I.1.

Table I.1: LIST Items and Loadings for Nine Factors

Organizing, Cronbach's α = .814		Loading
orga1	I make charts, diagrams and graphics in order to have the subject matter in front of me in a structured form.	.312
orga2	I compile short summaries of the most important contents as a mnemonic aid.	.689
orga3	I go over my notes and structure the most important points.	.626
orga4	I try to order the subject matter in a way that makes it easy for me to remember.	.340
orga5	I compile a summary of the main ideas out of my notes, the script or other sources.	.726
orga6	I underline the most important parts in my notes or in the texts.	.336
orga7	For bigger amounts of subject matter I find an arrangement that mirrors the structure best.	.601
orga8	I assemble important terms and definitions in my own lists.	.522
Elaborating, Cronbach's α = .766		
ela1	I try to find connections to other subjects or courses.	−.372
ela2	I think of practical applications of new concepts.	−.563
ela3	I try to relate new terms or theories to terms or theories I already know.	−.547
ela4	I visualize new issues.	−.616
ela5	In my mind I try to connect newly learnt facts to what I already know.	−.520
ela6	I think of practical examples for certain curricular facts.	−.563
ela7	I relate what I am learning to my own experiences.	−.530
ela8	I wonder if the subject matter is relevant for my everyday life.	−.334

LIST Items and Loadings for Nine Factors, continued 1

Repeating, Cronbach's α = .726		Loading
repeating1	I imprint the subject matter from the lecture on my memory by repeating it.	.194
repeating2	I read my notes several times in a row.	.233
repeating3	I learn key terms by heart in order to remember important facts better in the exam.	.379
repeating4	I commit a self-compiled compendium to memory.	.297
repeating5	I read a text and try to recite it at the end of each paragraph.	.392
repeating6	I commit rules, technical terms or formulas to memory.	.490
repeating7	I learn the subject matter by heart using scripts or other notes.	.590
Effort, Cronbach's α = .757		
effort1	Whenever I have planned a certain workload, I make an effort to master it.	.402
effort2	I make an effort even though the subject matter may not suit me well.	.491
effort3	I do not give up even though the subject matter is very difficult and complex.	.470
effort4	I work late at night or at the weekends if necessary.	.283
effort5	It usually does not need much time until I decide to start working.	.475
effort6	Before exams I take the time to go over all the subject matter again.	.350
effort7	I take more time for learning than most of my fellow students.	.482
effort8	I work until I am sure to pass the exam well.	.400

LIST Items and Loadings for Nine Factors, continued 2

Attention, Cronbach's α = .858		Loading
attention1	When I am learning I notice that my thoughts tend to stray.	.601
attention2	It is difficult for me to concentrate.	.708
attention4	When learning I am lacking in concentration.	.763
attention5	I am easy to distract when learning.	.726
attention6	My concentration does not last very long.	.751
Time Management / Planning, Cronbach's α = .790		
time1	I work according to a schedule.	.541
time2	I decide on the times for my learning.	.592
time3	I fix the hours I spend daily on learning in a schedule.	.630
time4	Before each study period I appoint the duration of my work.	.612
planning1	I try to consider beforehand which areas of certain topics I have to study and which I do not have to study.	.196
planning2	I decide in advance how much subject matter I would like to work through in this session.	.389
planning3	Before starting on an area of expertise, I reflect upon how to work most efficiently.	.363
planning4	I plan in advance in which order I want to work through the subject matter.	.354
Learning Environment, Cronbach's α = .700		
environ1	I work in a place that makes it easy to concentrate.	.343
environ2	I design my work environment in a way that I am distracted as little as possible.	.307
environ3	When learning I always sit at the same place.	.256
environ4	When studying I make sure that I can work uninterrupted.	.467
environ5	My workplace is designed in a way that makes it easy to find everything.	.522
environ6	At my desk I have the most important papers within reach.	.558

LIST Items and Loadings for Nine Factors, continued 3

Peer Learning, Cronbach's α = .783		Loading
peer1	I work on tasks together with my peer students.	.694
peer2	I take my time to discuss the subject matter with other students.	.656
peer3	I compare my notes with my peer students'.	.246
peer4	I make other students ask me questions on the subject matter and ask them questions too.	.310
peer5	I turn to help from others when I have serious problems in understanding something.	.541
peer6	When I am not sure about something I ask a fellow student for advice.	.742
peer7	If I find considerable gaps in my notes, I turn to fellow students.	.554

Using Reference, Cronbach's α = .765		
reference1	I search for explanatory material if certain facts are not completely clear.	.604
reference2	Whenever I do not understand a technical term, I look it up in a textbook or on the Internet.	.575
reference3	I look for missing information in different sources, e.g. the Internet, textbooks, or journals.	.734
reference4	When my notes are incomplete I use additional sources.	.628

J Graphs and Statistics for Paired Samples

This appendix contains additional figures for section 5.8.

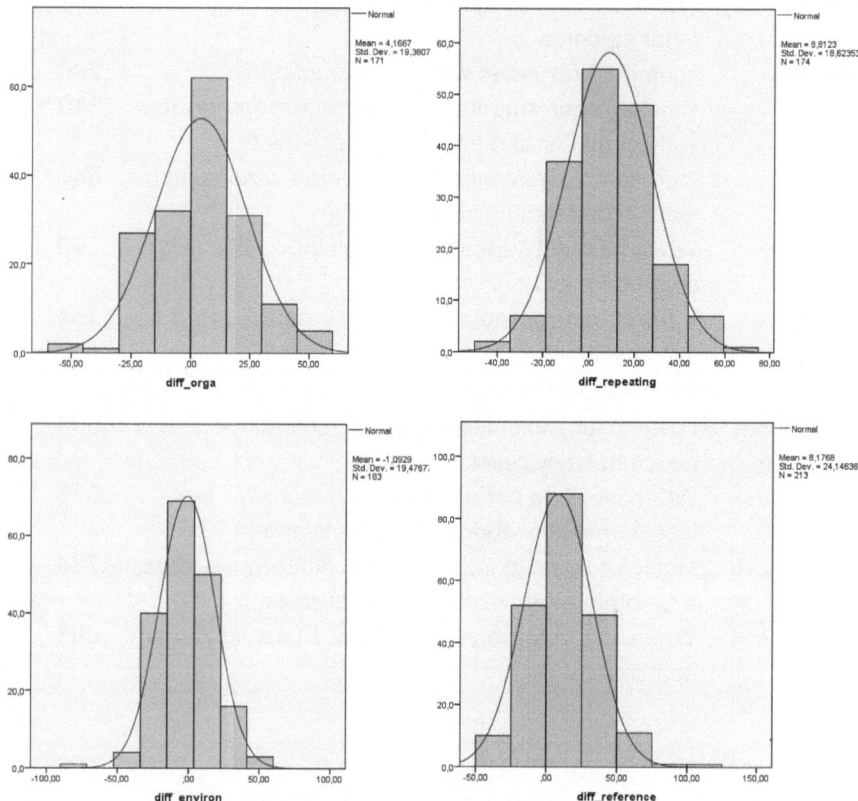

Figure J.1: Histograms for *Diff Organizing*, *Diff Repeating*,
Diff Learning Environment, and *Diff Using Reference*

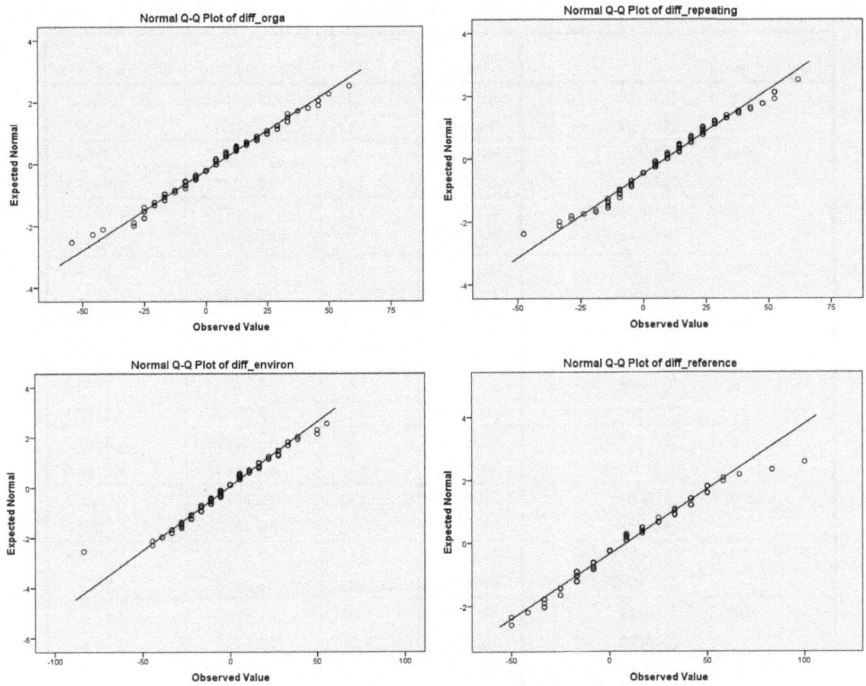

Figure J.2: QQ-Plots for *Diff Organizing*, *Diff Repeating*,
            Diff Learning Environment, and *Diff Using Reference*

Paired Samples Statistics

MP2-Teilnahme			Mean	N	Std. Deviation	Std. Error Mean
,0	Pair 1	@1orga	51,5467	132	21,03531	1,83089
		@2orga	46,7803	132	22,96413	1,99877
	Pair 2	@1ela	58,4341	124	16,38019	1,47098
		@2ela	47,0766	124	17,81273	1,59963
	Pair 3	@1repeating	47,0977	137	18,55705	1,58544
		@2repeating	38,3733	137	17,00111	1,45250
	Pair 4	@1effort	63,8105	124	17,15385	1,54046
		@2effort	58,0309	124	18,58971	1,66941
	Pair 5	@1attention	58,0365	146	26,21652	2,16969
		@2attention	44,4749	146	24,74835	2,04819
	Pair 6	@1time_planning	41,0714	147	18,74841	1,54634
		@2time_planning	42,8288	147	19,58462	1,61531
	Pair 7	@1environ	62,6437	145	19,87490	1,65052
		@2environ	64,1762	145	18,29648	1,51944
	Pair 8	@1fellow	59,1631	132	18,95763	1,65005
		@2fellow	57,3954	132	18,45042	1,60590
	Pair 9	@1reference	77,0964	159	18,93732	1,50183
		@2reference	67,3480	159	24,98186	1,98119
1,0	Pair 1	@1orga	54,1667	38	21,09075	3,42137
		@2orga	52,9605	38	22,44671	3,64134
	Pair 2	@1ela	54,1667	36	18,12107	3,02018
		@2ela	43,6343	36	22,26032	3,71005
	Pair 3	@1repeating	53,1746	36	17,83559	2,97260
		@2repeating	44,1799	36	15,25556	2,54259
	Pair 4	@1effort	66,8803	39	17,88192	2,86340
		@2effort	64,4231	39	18,75234	3,00278
	Pair 5	@1attention	51,5873	42	26,30072	4,05829
		@2attention	42,8571	42	25,59399	3,94924
	Pair 6	@1time_planning	45,6081	37	20,19738	3,32043
		@2time_planning	50,4505	37	23,04852	3,78915
	Pair 7	@1environ	66,6667	37	19,51047	3,20750
		@2environ	67,4174	37	19,99495	3,28715
	Pair 8	@1fellow	56,5760	42	20,80197	3,20981
		@2fellow	57,9365	42	21,44899	3,30965
	Pair 9	@1reference	78,3019	53	20,36916	2,79792
		@2reference	74,8428	53	21,83432	2,99917

Figure J.3: Statistics for Paired Data, Separately for MP^2-Math/Plus Participants (1,0) and Non-Participants (,0);
@1 for pre scores, @2 for post scores